图书在版编目（CIP）数据

中国当代建筑大系——绿色建筑 / 杨维菊编；常文心译. -- 沈阳：辽宁科学技术出版社，2013.3
ISBN 978-7-5381-7792-3

Ⅰ. ①中… Ⅱ. ①杨… ②常… Ⅲ. ①生态建筑－建筑设计－作品集－中国－现代 Ⅳ. ①TU206

中国版本图书馆CIP数据核字（2012）第286777号

出版发行：辽宁科学技术出版社
　　　　（地址：沈阳市和平区十一纬路29号　邮编：110003）
印 刷 者：利丰雅高印刷（深圳）有限公司
经 销 者：各地新华书店
幅面尺寸：235mm×305mm
印　　张：32
插　　页：4
字　　数：50千字
印　　数：1～1700
出版时间：2013年 3 月第 1 版
印刷时间：2013年 3 月第 1 次印刷
责任编辑：陈慈良　杨子玉
封面设计：杨春玲
版式设计：杨春玲
责任校对：周　文
书　　号：ISBN 978-7-5381-7792-3
定　　价：228.00元

联系电话：024-23284360
邮购热线：024-23284502
E-mail: lnkjc@126.com
http://www.lnkj.com.cn
本书网址：www.lnkj.cn/uri.sh/7792

CONTEMPORARY ARCHITECTURE IN CHINA

中国当代建筑大系

绿色建筑
GREEN
ARCHITECTURE

杨维菊/编　常文心/译
Edited by YANG Weiju
Translated by Catherine Chang

辽宁科学技术出版社

FOREWORD

To Create Green Architecture with Chinese Characteristics

Nowadays, in front of the challenge of global climate changes, green architecture is the most important solution. In the process of reducing the ratio between GDO and carbon emission to 40%-45% compared to 2005, green architecture holds great responsibilities. In the process of developing low-carbon eco city, green architecture will make major contributions. In this sense, the efforts we've made for the promotion of green architecture are both beneficial for this generation and our future generations. [1]

With the promotion of sustainable development all over the world, green architecture gains more attention from various countries. Recently, some forward-looking Chinese architects are carrying out relative researches and explorations in practice. They've expanded the research of green architecture into a broader field.

There are many excellent foreign cases in these practices, including Tjibaou Culture Centre by Renzo Piano and UNMO by Kenneth Yeang. The Indian architect Charles Correa insists that "forms follow climate" and creates architectural design on the basis of climatology, creating unique architectural forms. His master work Kanchanjunga Apartments is developed from India's hot and humid climate with careful considerations of prevailing wind direction and landscape orientation, in combination of Indian heritages and local architectural elements. (Image 1) According to the local climate, the best orientation is west, which could help enjoy the cool breeze from Arab. Therefore, each apartment occupies two storeys or partially two storeys, with a two-storey corner terrace garden. The small windows could avoid the attack of sunlight and monsoon rain, while the two-storey terrace provides the inhabitants with sea wind and view of the nearby Bombay harbour. In addition, the upper level of each apartment has a small balcony open to the terrace garden. In this project, the architect perfectly solved the principal contradictions of monsoon, western sunlight exposure and landscape. The openings and colours on the façade imitated the methods of Le Corbusier and attached to each unit its identity. As Correa's only high-rise residential building, Kanchanjunga Apartments highlights his comprehensive ability in spatial treatments, excavation of traditional architectural essences and climate- and landscape-oriented design. At that time, the architectural form is "both chic and India inspired". [2]

Today, we come to realise the importance of sustainable development and green architecture generally enters into our view. Modern architecture no longer highlights the priority of function issues only; it requires more in architectural forms, living comforts and environmental sustainability. Architectural types, materials and methods are enduring fierce revolution. The goal of green architecture is to facilitate the architecture to make full use of the climate and adapt to it. The architecture should act according to local circumstances and optimises the environmental resources. We should improve and create comfortable living and working environment through plan, design and environmental configuration, achieving architectural "conservation of energy, land, water and material", in order to create healthy and comfortable exterior and interior environment and reduce the negative effects for the environment.

The performance of green architecture depends on the design process of the architect. The energy consumption is not only related to the thermal insulation of the building's envelope, but also the architectural form, floor plan, spatial organisation, elevation form, architectural construction, materials and master plan of the architecture complex. Therefore, once the enlarged preliminary design is finished, the energy consumption is determined. If the architect could design according to the relationship between the architecture and the local climate and follow the principles of green architecture, he will be able to design an energy-saving architecture. [3]

Now, Chinese architectural professionals have published numerous articles about green architecture, which play an active role for the promotion of sustainable development and development of green architecture. However, most of the articles are reports on theory research, design principles, ecological references of green architecture design and introduction of advanced experiences of green architecture aboard. They lack of practical engineering practice model, computer simulation and analysis of environmental effects. On one hand, we indeed should improve our theories about green architecture research; on the other hand, we need more practice to conclude, create, explore and develop green architecture with Chinese characteristics.

Contemporary Architecture in China: Green Architecture is a book worth reading, which includes more than 20 successful projects. These projects are all recent projects in China, most of which are completed between 2010 to 2011, including energy technical research centre, commercial complex, technopark management centre, financial tower, international cruise terminal, ZED pavilion, office building, laboratory building, school, residence and industrial park.

Among them, the creative Vanke Centre, Shenzhen is the first building to be certificated as LEED platinum by USGBC; committed to be the best green commercial building in China, Park View Green, Beijing won Mipin Asia 2010 (Best Green Building in Asia) with its unique form and ecology feature; enjoying the the ocean view and night view of the harbour, the design of Shanghai International Cruise Terminal combines modern architectural technology and materials, awarded as pioneer work of Shanghai sustainable development; China National Offshore Oil Corporation has won 2007 AIA Hong Kong – Architecture Merit Award and Sustainability in Design Award; Awarded 2011 Jiangsu Excellent Architecture Design – First Prize, R&D building of Wuxi Suntech Power Company is a perfect combination of solar energy and architecture. Many other green projects in this book have got relative certifications or are proceeding green certifications.

前　言

创造具有中国特色的绿色建筑

今天，在全球气候变化的挑战面前，绿色建筑是最重要的应对领域；在实现到2020年我国单位国内生产总值与二氧化碳排放比2005年下降40%~45%目标的征途上，绿色建筑肩负重任；在建设低碳生态城市的进程中，绿色建筑将作出重大贡献。从这个意义上讲，我们为推广绿色建筑所做的一切努力将无愧于现在，也必将无愧于后人。[1]

随着可持续发展观念在全球的推广，绿色建筑在世界各国受到广泛的关注。近年来，国内外一些有远见的建筑师在实践中已经开展了相关的研究和探索，他们将绿色建筑的研究拓展到更加广阔的领域。

在这些实践中，国外做得比较好的建筑师不少。如皮亚诺设计的吉巴欧（Tjibaou）文化中心和杨经文设计的梅纳拉UNMO大厦等。而印度建筑师查尔斯·柯里亚所坚持的"形式追随气候"把建筑设计建立在气候学基础之上，从而诞生出独特的建筑形态。其著名代表作干城章嘉公寓楼就是根据印度高温湿热的独特气候环境特点，充分考虑建筑所在地的主导风向和景观朝向，结合对印度传统和乡土建筑的设计要素，合理提炼而成的（对页图）。根据当地气候特点，最好的朝向是西向，而且有来自西边阿拉伯的凉风。因此，公寓楼每户占两层或局部两层，并有一个两层挑空的转角平台花园。房间的窗较小，可免受日晒和季风雨的侵袭，但挑空平台使住户充分享受到附近孟买港的海风与海景。每户上层的房间还有小阳台向平台花园开敞。建筑师在这个项目里完美地解决了季风、西晒和景观三个主要矛盾。建筑立面的开洞和色彩借鉴了柯布西耶的手法并赋予各单元以识别性。干城章嘉公寓楼作为柯里亚建成的唯一的一座高层住宅，凸显了其处理空间、挖掘传统建筑精华、针对气候和景色进行设计的综合能力，建筑样式在当时"既新潮，又富有印度风格"。[2]

当前，世界范围内人们开始意识到建筑"可持续发展"的重要性，以"绿色建筑"为目标的建筑正逐渐进入人们的视野。现代建筑已不再满足原来的仅仅优先解决功能问题，而是对建筑造型，居住的舒适性，环境的可持续性提出更高要求，建筑形式、建筑材料、建造方式等都在发生深刻变革。如今，绿色建筑设计追求的目标是使建筑可以更充分利用和适应气候条件，做到因地制宜，合理利用环境资源，以规划、设计、环境配置的建筑手法改善和创造舒适的居住、工作环境，达到建筑的"节能、节地、节水、节材"，创造出健康舒适的室内外环境，降低对环境的负面影响。

我们说，绿色建筑性能的好坏与建筑师的创作设计过程有着直接关系，因为建筑能耗不仅与建筑围护结构的保温、隔热性能有关，还与建筑形体、平面布局、空间组织、立面形态、建筑构造、材料选用及建筑群体规划布置等都有着密切的关系。因此，在建筑的扩初设计完成后，建筑物的耗能指标就已确定，如果建筑师在方案设计的每一环节都考虑到建筑与地域气候要素的关系，并能遵循绿色建筑的准则，就可能设计出运行能耗低的建筑。[3]

目前，我国建筑界发表了许多关于绿色建筑的文章，这些文章对宣传可持续发展的思想，推动绿色建筑的发展起到了一定的积极作用，但绝大多数的文章都停留在绿色建筑设计的理论框架、设计原则及生态学理论对建筑的指导，在国外先进绿色建筑的经验介绍方面，缺乏实际工程实践模式、计算机模拟、环境效果的测试分析，关于绿色建筑理论研究是应该加强，但更多的应在工程实践中进一步地总结、创新和摸索，研究适合中国国情的绿色建筑。本书《中国当代建筑大系——绿色建筑》是一本非常值得一看的好书，它有20多个成功的案例介绍。这些案例都是近几年国内建成的新建筑，其中2010年-2011年的工程占多数。它们包括：能源技术研究中心、商业综合体、科技园管理中心、金融大厦、国际港客运中心、零能馆、办公楼、实验楼、学校、住宅、工业园改建等不同类型的项目。

其中，位于深圳的"万科中心办公楼"，设计构思立意新颖，是中国第一座由美国绿色委员会认证的白金级绿色建筑；北京芳草地是打造中国最绿色的节能商业综合体，造型独特，生态环保，在2010年获亚洲最佳绿色建筑奖；而上海国际港客运中心独特的设计将海景、夜景收于眼下，将现代建筑技术与材料完美结合，被评为上海可持续建筑发展的先驱作品；中国海洋石油总公司大楼被美国建筑师协会香港分会评为2007年环保建筑设计优异奖；无锡尚德太阳能电力有限公司研发大楼，气势磅礴，雄伟壮观，是太阳能与建筑一体化的完美展现，获得2011年江苏优秀建筑设计一等奖。书中所介绍的其他绿色工程也已通过或正在申请相关的认证。

从本书优秀案例的介绍中，我们领略了近几年中国绿色建筑的发展趋势、设计理念及材料、技术和艺术的完美结合。它们为中国今后的建筑创作提供了极具价值的经验和技术支持。总结归纳这些优秀案例的特点，分析它们的优势所在，有助于寻求我国当前可以借鉴的绿色建筑的途径。

1. 全新的设计理念

一个好的、成功的建筑，其设计思想和理念是非常重要的，它是整个作品的灵魂。设计重在创作的思维，就绿色建筑而言，它可以被理解为是传统建筑设计的延伸和深化，它与传统建筑设计的区别在于其是基于整体的绿色化和人性化设计理念，并在此基础上进行综合创新和系统设计。具体来说，就是在满足功能和性能的基础上，在整个建筑的全过程中，尽可能少的利用有限资源，实现资源最大的利用率，尽可能以较少的资源和较低的环境污染，为人们创造一个和谐、健康、高效的可利用空间，实现人与自然的和谐共处。

以科技为先，以创新的理念和思想方式来进行设计，用绿色建筑设计的理论框架、设计原则及生态学理论对建筑进行指导。在设计中必须将自然因素考虑到其中，并将其融入周围环境，使生态环境和人文环境能有效地融合在一起。在设计中考虑以本地文化为主线，充分运用原材料和建筑资源，在尊重自然的前提下，构筑有地方特色的建筑物。[4]

案例"巨型花架"苏州生物科技园，它的设计目标就是要创造一个可持续发展的生态智能化科技园区，其中管理中心最基本的设计意图是创建开放式的、生态化的办公空间。两座平行的办公建筑之间自然形成了一个园林广场，从而将中央公园引入管理中心，通过圆形设计元素从中央公园一直延续到管理中心的外立面和中庭，整个管理中心犹如一个可呼吸的"绿肺"。同时，建筑以及建筑之间的园林上覆盖着技术简单、成本低廉、效能良好的遮阳棚，解决了园林广场和建筑的遮阳问题。从中可以看到当地传统的建筑和园林关系被继承保留并革新性地再现。

"万科中心"的设计理念也是别具一格的。整个建筑看起来就像浮在一片退潮的海水上，在建筑和景观设计中采用一些全新的可持续措施。建筑的绿色屋顶上配有太阳能光电板，并将竹子等本地材料应用在门、地面和家具中。建筑的玻璃外立面通过多孔的百叶窗进行遮阳和防风。建筑具有防海啸性能，在开阔的公共景观中打造了一个可渗透的微环境。

2. 空间与形体的塑造

在当代建筑迈向"绿色与可持续发展"的道路上，技术手段扮演着先锋的重要角色。对于未来建筑的发展方向，以空间和形体塑造为特征的传统建筑学正面临着巨大的挑战。我们不可否认绿色技术在建筑中所发挥的重要作用，但也要清醒地看到，仅依赖技术的推动力来解决问题将会面临成本过高、负面效应上升的不利局面，因此，单纯的技术探索能否持续高效地发挥作用，还需要我们进一步的深入研究。

现今，我们也看到了由于建筑功能和使用效率的要求，现代建筑的体量在不断增大，因此出现了越来越多规模庞大的建筑综合体。然而，建筑

From these excellent projects, we could know about the development tendency, design concepts, and the perfect combination of materials, technologies and art of Chinese green architecture in recent years. They've provided practical experiences and technology supports for our future architectural creations. Concluding these successful projects and analysing their advantages is helpful for us to seek the references for green architecture.

1. Brand New Design Concepts

For a good and successful architecture, the design idea and philosophy are very significant, which are the soul of the architecture. Simultaneously, design emphasises creative thinking, especially for green architecture; it could be understood as the expansion and improvement of traditional architectural design. Their differences lie in the integral creative system design based on the overall green and human design concept. Specifically, based on the fulfillment of function and performance, the whole process of construction tries to utilise the limited resources to realise the maximum use ratio. The process tries to use less resources and makes lower environmental contamination to create a harmonious, healthy and effective space for us, achieving the harmonious co-existence between human and nature.

Relying on science and designing with creative concepts and thinking, design principles and ecological theories of green architecture will guide the architectural design. We must consider the natural elements in the design and merge the building into the surroundings, effectively blending eco and human environment. Taking local culture as the main line and making full use of materials and architectural resources, on condition of respecting nature, designers will construct architecture with local characteristics. [4]

"Giant Veranda" – Suzhou Industrial Park aims to create a sustainable eco intelligent science park. The basic design intention of administration centre is to create an open eco office space. Two parallel office buildings form a natural landscape plaza in between. Therefore, the central park is introduced into the administration centre. The central park continues to the administration centre's façade and atrium through circular elements, making the whole centre look like a breathing "green lung". Meanwhile, the landscape garden is covered with inexpensive yet effective shading grids with low technology, which solves the shading problem of the plaza and the buildings. The local traditional architecture and garden design has been inherited and renovated in this project.

Vanke Centre's design concept has a unique style. The whole office building seems to float on the fall of tide. The project applies some new sustainable methods in architectural and landscape designs. The green roof is installed with solar photovoltaic panels while the doors, floors and furniture of Vanke Centre all uses local materials such as bamboos. The glazing façade is shaded from the sunlight through porous louvres. The building is tsunami-proof and creates a penetrable micro climate in an open public landscape.

2. Creation of Spaces and Forms

In the progress of contemporary architecture to become "green and sustainable", technology plays a pioneer role. In the future architecture's development, traditional architecture which features the creation of spaces and forms is facing a great challenge. We couldn't deny the important role of green technology in architectural design. However, we should also be aware that if we only rely on the technology to solve the problems, we will face high cost and negative effect. Therefore, we need further research about whether pure technology could play a continuous effective part.

Today, because of the requirement of architectural function and use efficiency, the volume of modern architecture is increasingly large and there appears more and more giant architectural complexes. However, with the increasing high requirement for ventilation, the control of form factor is no more a problem. The oversize depth becomes a barrier for improving the architectural environment. Specific to this issue, the projects in this book demonstrate how to lead the flow of interior air better and to strengthen the permeation of natural light, which advocates a design thinking – taking natural conditions as priority to satisfy the requirements of interior comfort. Parkview Green Beijing and Shanghai International Cruise Terminal and some other projects all have distinctive characters in form design.

Completed in 2010, Parkview Green is one of the most green and efficient commercial complexes in China. It innovatively envelopes 230,000 square metres of commercial, office and hotel spaces in a shield constructed of steel, glass and ETFE cushions. In the process of construction, the project overcame numerous technical and legal challenges and finally became Beijing's landmark due to its distinctive form and green technologies. Aiming to protect the preferable daylight and natural ventilation of the surrounding residents, in the unshielded sunlight exposure and under the strict restriction of building height, the architect designed a pyramid building. In order to create a good thermal environment, the whole building is 9 metres lower than the ground, so that the sunken garden will introduce fresh air and the heat effect will extract the interior hot air out, providing each level with natural ventilation. Compared with traditional architecture, Parkview Green could reduce energy consumption by 16% in summer, and 83% in winter. Besides, the 235-metre-long footbridge supported and hang by steel inside the building links the cross corners of the ground floor, which becomes a public road for people to cross the downtown. Passengers on the bridge will experience a unique urban space which is full of modern art works and convenient for street activities. Through the application of a series of design methods and technologies, Parkview Green creates a distinctive urban experience in Beijing CBD.

Known as a "crystal palace" by the Huangpu River, Shanghai International Cruise Terminal is a new destination for Shanghai in the 21st century. In consideration of future development, the project requires 50% of the building to be underground. The designer focuses on how to treat the relationship between "under world" and the architecture rising out of it. The layered undulating landscape design introduced curtain wall to become a second skin for the building, avoiding hard light for the south-facing commercial office space. In the meantime, the designer set exterior balconies overlooking Huangpu River between the two curtain walls. What is smarter is an intriguing gap in the middle – a glazed table top supports pods on cables, hovering over a public performance space below. The pods

对于通风散热的要求不断提高，体形系数的控制不再是问题的关键，过大的进深反而成了改善建筑环境的障碍。针对这个问题，书中的实践案例集中展示了如何更好的引导室内空气的流动、加强自然光的渗透，倡导了一种优先利用自然条件来满足室内环境舒适度要求的设计思路。如北京芳草地商业综合体、上海国际港客运中心等几个案例，在可见形体的塑造上都各有特色。

2010年建成的北京芳草地项目是目前国内设计的最绿色的节能商业综合体之一。它标新立异，将占地23万平方米的商业、办公和酒店用一个由钢铁、玻璃和ETFE构成的微环境表皮包裹起来。在整个建造过程中，该建筑克服了众多技术和法规上的挑战，最终以独特可辨的造型和绿色设计成为北京地标。它的设计思路就是要保护周边居民具有较好的自然采光和通风，在毫无遮挡的日光角度及严格的建筑物高度限制下，建筑师将建筑形体呈现为一个金字塔式的造型。为了创造良好的热舒适环境，整个建筑设计低于地面9米，这样就可以从下沉花园引入新鲜空气，并在热效应的作用下将室内的热气从顶层拔出，从而使建筑的每一层都能享受到自然通风。与传统建筑相比，芳草地项目在夏季可节能16%，而在冬季节能则可达83%。此外，在建筑内部之间设有一条235米长、由钢铁支撑和悬挂的人行桥，并斜线连接首层的对角，成为人们穿越城市区的一个公共道路，使经过桥上的人们感受到一个充满现代艺术气息又便于街道活动的、与众不同的城市空间。通过这一系列的设计手法和技术应用，芳草地项目打造了北京CBD中心区独特的城市体验。

上海国际港客运中心案例被称为黄浦江边的又一个"水晶宫"，也成为21世纪上海新的一景。考虑到今后的发展，在项目中，要求建筑50%的面积需作为地下空间，设计师的构思重点是如何处理好"地下世界"及地上的建筑。造型上，层层线性起伏的景观设计引入幕墙设计，成为大楼表面的第二层幕墙，避免南向商业办公空间受强光影响。同时，在两层幕墙之间设置了供人们俯视黄浦江的室外阳台。更巧妙的是在其中两幢大楼之间设置了里面悬挂着数个吊舱的"玻璃桌子"，而吊舱悬浮于一个公共演出的空间之上，分设咖啡馆、酒吧和餐厅，把建筑空间的设计和多元娱乐需求融为一体。

3. 技术集成策略

绿色建筑技术并不独立于传统建筑技术，而是用"绿色"眼光重新审视传统建筑技术，是传统建筑技术与相关学科的交叉与组合。

现今，节能环保的绿色建筑其立面和造型已不仅是传统意义上的围护结构，而是被赋予了各种节能技术的新型墙体。建筑师应该有多学科交叉的设计和技术思维，如果囿于原有的知识结构，仅着眼于建筑形式设计本身，则远不能解决未来的环境与城市问题。另外，绿色节能建筑在设计中不但要注重地域特色，还应考虑与当地气候条件协调，并要优先利用自然通风、天然采光、建筑遮阳等被动式节能技术，以集成的思维来使用各种低能耗技术。

本书有几个建筑案例就具有代表性。可持续技术研究中心楼，其主要功能是为员工和研究生提供一个专业研究的实验室、办公室、研讨教室等空间。它展示的是可持续建筑与节能内部环境控制等前沿技术，通过使用生态技术，减少能源消耗；利用当地材料和可再生能源，减少对环境的影响。技术研究中心通过五个环境设计策略来进行工作：①高性能的围护结构；②暴露的蓄热体；③太阳光控制；④塔楼的自然通风；⑤实验室和加工间的管道通风。在能源获取方面，部分计算机、照明设备等用电都利用可再生能源转化来获取，如地源热泵、太阳能吸收制冷、太阳能光电板等。该项目计划不使用传统的加热和冷却系统，而且民用能源需求由可再生资源来满足，降低碳排量，力求通过众多生态技术的应用来证明可持续能源技术将为未来的低碳经济做出更大的贡献。

坐落在清华大学校园里的中意环境节能楼，通过造型和外立面的设计来控制外部环境，从而获得最佳的内部环境。建筑师为了尽可能减少冬季寒风对大楼的侵袭，北立面采用完全不透明而且保温性能良好的材料，东立面和西立面采用了双层幕墙以控制光线和日光的直射，使整个办公空间既可以控制阳光，又实现了最佳的采光效果。南立面同样是双层玻璃幕墙，上装有光伏电板，呈梯形悬垂分布在东南立面的表面上，同时起到了遮挡太阳对玻璃幕墙照射的作用。大楼面向天井的双翼内侧采用了双层单幕墙系统，在外层安装了由倾斜角度各不相同的反射玻璃构成的玻璃百叶，避免室内被强光直射，同时改善自然光照明的效果。另外，建筑南侧的露台和悬臂式结构，能为下层露台遮阳，也是太阳能光伏发电系统的支架。大楼设有内庭花园，布置得错落有致，配以树木、流水和花草，庭内的浅水池（300毫米深）和绿化区共同构成具有意大利花园的风格。

无锡尚德太阳能电力有限公司研发楼将建筑造型和节能技术结合，在创作中充分挖掘可再生能源的潜力，推进太阳能光伏电板在建筑中规模化的应用，实现建筑创新，使研发楼建筑超大、超简洁的外观与室内空间的丰富灵动形成强烈对比，成为建筑的最大亮点和特色。大片的太阳能光电板墙以一种简洁有力的形象拔起，极富视觉冲击力。主体厂房部分外墙采用不锈钢架和绿色植物组成的绿色外衣将建筑包围，竹状的绿色构件，体现出江南特有的清灵韵味。建筑采用自身产品——太阳能光电板，取代传统的玻璃幕墙，充分体现了可再生能源利用和建筑外形设计的统一性。研发楼除了太阳能的利用，还有地源热泵系统及其他技术的运用，使研发楼最终实现"零能耗"的目标。

从本书可看到，大多数的实际项目案例，都有一定的个性、代表性和设计特点。它们应用一些绿色的技术措施，使建筑达到节能、低能耗和绿色环保的目的，如：多种遮阳形式、自然通风、地源热泵的利用、节能的围护结构和低辐射玻璃的使用等。同时在设计风格、建筑造型、反映时代特色上也别具一格，注意朝向，平面布局自由灵活，有些建筑通过空中花园的设置，以及景观、园林、街景的配合，使建筑更加新颖、更加现代、更富魅力。

备受关注的上海世博会零碳馆展现了中国式街区的未来场景，造型简洁、现代，屋顶体现了太阳能与建筑一体化的设计理念，并将绿化屋顶和阳台巧妙地融合到建筑的造型之中，使人感到设计新、材料新、技术新。这种创造性的建筑材料源于中国，而且造价低廉，并已形成本土的供应链体系，为中国的零碳城市化奠定了基础。零碳馆的样板房建筑采用了无动力通风系统和太阳能除湿系统，缓解了空调的巨大能耗，在复式住宅中，房间和阳台均朝南以获取最大化采集太阳能。建筑内使用的电能均由屋面太阳能光伏板提供，并设计栈桥将建筑单元连接起来，以促进邻里互动。在墙体、地板和屋顶采用内部控温，空间采用双倍层高，更适合商业和办公，同时，利用北向开窗避免了过高的室温。另外，城市化零碳系统在节约资源的同时还为多数居民创造了更高质量的生活环境。

4. 结语

随着社会快速发展，我国新技术、新材料、新产品不断出现，绿色建筑已成为建筑走向可持续发展的必然选择。按照可持续发展要求，建筑设计的复杂性和专业性增强，从建筑整体宏观的角度来设计建筑，已成为改变建筑师设计理论的新要求。

同时，绿色建筑的控制将在绿色建筑评价的基础上更全面地对建筑引进系统化思考，它包括建筑师的定位以及设计工作中做好各专业协调整合的问题。近年来，由于宣传力度的加强和人们认识的提高，国内绿色建筑的发展态势较好。国家及各地方政府积极推广绿色建筑，其效果是显著的。我国的绿色建筑工作取得了飞速发展，绿色建筑技术体系不断完善，绿色建筑评价标识项目数量这几年显著增多，整个绿色建筑行业已经显

contain cafés, bars and restaurants, combining spatial design and entertainment requirements together.

3. Technology Integration Strategy

Green architectural technology isn't independent of traditional architectural technology, but takes a new look on it. It is the intersection and combination of traditional architectural technology and its related subjects.

Nowadays, the façade and form of energy-saving and environment-protecting green architecture is no longer the traditional envelope, but new-type wall with various energy-saving technologies. An architect should possess multi-disciplinary design and technology thinking. Pure architectural form design and closed knowledge system cannot solve future conflicts between environment and urbanization. In addition, green architecture should not only emphasise regional characteristics, but also consider local climate and utilise passive energy-saving technology including natural ventilation, natural lighting, solar shading, using various low-energy consumption technology with integral thinking.

This book includes several representative cases. Centre for Sustainable Energy Technologies provides staff and researchers with professional laboratories, offices and seminar rooms. It demonstrates state-of-the-art techniques such as sustainable architecture and energy-efficient internal environmental control, and how to reduce energy consumption through eco-technology. It uses local materials and renewable energy sources to reduce the environmental impacts. The centre works through five environmental design strategies: High Performance Envelope, Exposed Thermal Mass, Solar Control, Natural Ventilation to Tower and Piped Ventilation to Laboratory & Workshop. In term of energy acquisition, part of the power comes from renewable energy, such as ground-source heat pump, solar absorption cooling system and photovoltaic panels. The project abandons traditional heating and cooling system and the energy demands could be met by renewable resources to reduce the carbon emission. The project tries its best to demonstrate the contribution that sustainable energy technologies can make to the low carbon economy of the future.

Sino-Italian Ecological and Energy Efficient Building of Qinghua University controls the external environment through the design of its shape and of its envelope in order to optimise the internal environmental comfort conditions. In order to minimise the invasion of winter wind, the north façade uses solid materials with good thermal insulation. The east and west façades use double glazed wall to control light and direct sunlight. Therefore, the whole office space could both control the sunlight and achieve the best lighting effect. The south façade's curtain wall is installed with photovoltaic panels, which are vertically arranged in trapezium on the southeast façade, shading the sunlight at the same time. Two wings facing the patio use double curtain wall system. Reflecting and semi-reflecting louvres will allow for sunshine to penetrate in the rooms and avoid direct sunlight. Besides, the south terrace and cantilever structure, which is also the supports of photovoltaic panels, will provide shade for the terraces on the lower levels. The building has a well-proportioned internal garden, completed with trees, stream and various plants. The pool (300mm in depth) and green area constitute an Italian style garden.

R&D Building of Wuxi Suntech Power Company combines its architectural form with energy-saving technologies and excavates potentials of renewable energy. It promotes the large-scale application of photovoltaic panels in architecture to achieve architectural energy creation. The giant and simple exterior makes a strong contrast with the rich and lighting interior space, which becomes the most impressive feature of the buiding. The solar photovoltaic panel walls rise from the earth in a concise and strong way. The main plant area uses stainless frame and green envelope to wrap the architecture. The bamboo green construction also makes the large-scale architecture demonstrate an inspiring charm of south region. The architecture uses Suntech Power's own product – Solar photovoltaic panel to replace conventional glass wall and to function as envelope to the main façade. The application of Ground Source Heat Pump, on the base of application of solar energy, further advances the renewable application in the architecture, finally achieving the goal of "Zero-energy" building.

In this book, most practical projects own some distinctive and representative design features. They apply some green technology to achieve an energy-efficient and green building. For example, various sunshade types, natural ventilation, ground source heat pump, energy-efficient envelope and low-emission glass. They are also distinctive in design styles, architectural forms and modern characteristics, paying attention to orientation and flexible layout. Through sky gardens and coordination between landscape, gardening and streetscape, some buildings look more novel, chic and charming.

The attracting Shanghai Expo ZED Pavilion unfolds a future scene of a Chinese block, with a concise and modern form. The roof demonstrates the integration of solar energy and architecture. The smart design to incorporate the green roof and balconies into the architecture is new in design, materials and technologies. All of the low-cost innovative building components were sourced in China, and ZED factories have established a supply chain to inform the longer term roll out of zero carbon urbanism. The zero carbon building utilises natural ventilation and solar dehumidification to relieve the large energy consumption of air conditioning. In the duplex apartment, both the rooms and balconies are facing south in order to get maximum solar collections. All the power used by the building is provided by the photovoltaic panels on the roof. All the unites are connected by bridges, helpful to the interactions between neighbours. The designers use internal temperature control in walls, floors and roofs. Double-height space is suitable for commercial and office uses. In the meantime, the north-facing windows avoid high temperature. The zero carbon system not only saves considerable resources, but also creates a better living environment for most inhabitants.

4. Conclusion

With the rapid development of society, new technologies, new materials and new products continue to emerge. Green architecture has become an inevitable option of sustainable development. According to the requirements of sustainable development, the complexity and professionalism of architectural design are reinforced. To design architecture in a macro view has become new requirements to change the architects' theories.

At the same time, the control of green architecture will introduce systematised thinking on the basis of green architecture evaluation, which includes the self-positions of the architects and the coordination between different disciplines in the design. Recently, with the enhancement of publicity and people's cognition, Chinese green architecture develops very well. The national and local governments are positively promoting the development of green architecture and the effect is remarkable. We have achieved rapid development in the field of green architecture and the technology systems are increasingly completed. Projects with green architecture label grow remarkably. The whole green architecture industry is obviously in a rapid development starting phase. From 2008 to 2010, we already have 113 projects with green architecture labels. From 2011 to 2015, China will face a grand tide of green architecture development and green architecture will develop and be promoted with great efforts, which is significant to solve China's energy and environmental issues. [5]

Here is China's guidance of green architecture: in term of energy saving, we highlight energy-saving design and the application of renewable resources, as well as the optimisation of positive energy-saving technology on the basis of passive energy-saving technology; in term of water saving, we highlight the recycle of water; in term of land saving, we highlight the efficient use of land; in term of material saving, we highlight the use of energy-saving materials and the full-decoration concept; in term of operation, we highlight the intelligent management and efficient operation. [6]

Today, green architecture has become a new trend in the development of architecture all over the world, especially in this age suffering from resource shortage. During these years, China has developed a systemised design strategy of green architecture from concept to practice. All kinds of new technologies and new materials have also pushed the development of green architecture. However, we must realise that we should not copy foreign technologies and strategies in China directly, but make a choice according to China's actual conditions and specific environmental and climate characteristics. In the meantime, we should use homemade materials to reduce the cost and give architectural design a new concept in a sustainable view. In order to achieve greening in architectural design, we must introduce integral design concept and pay attention to technology integration and optimisation, so that the technology effects could be maximised in architectural design. We should give full play to creation and initiation and depend on the coordination between multiple professions and disciplines. The study of system-theory should combine with engineering practices. After continuous explorations, tests and summaries, we will create green architecture with Chinese characteristics.

Southeast University
Yang Weiju

References:
[1] Qiu Baoxing. China's Green Architecture Prospect and Strategy Advices[J]. Architecture Science and Technolgy, 2011/06.
[2] Li Yuan, Qin Qin. From Kanchanjunga Apartments to View Charles Correa's Empty Spaces[J]. Shanxi Architecture, 2008/11.
[3] Liu Jiaping, Tan Liangbin. A Discussion of the Development of Green Architecture[J]. Urban Architecture, 2008/04.
[4] Wang Yuzhe. The Application of Green Architecture Concept in Architectural Design.
[5] Cheng Zhijun. A Review of China's Projects with Green Architecture Labels (2008-2010)[J]. Eco City and Green Architecture, 2011/01.
[6] Zhang Zhen. China's Green Architecture Status and Strategy[J]. China Science and Technology Information, 2008/06.

处于高速发展期的起步阶段。2008年-2010年这三年内，我国已有113个绿色建筑标识项目，在"十二五"期间，国内将迎来绿色建筑的发展大潮，绿色建筑将得到大力发展和推广，这对于解决中国能源与环境的问题有着非常重要的意义。[5]

目前，我国绿色建筑引导的方向：在节能方面，强调更高标准的节能设计、可再生能源的应用，以及主动节能技术在被动节能技术基础上的优化；在节水方面，强调水资源的循环利用；在节地方面，强调土地的高效利用；在节材方面，强调使用节能材料以及全装修概念；在运行方面，强调智能化的管理和高效的运行等。[6]

今天，绿色建筑已是世界建筑发展的新趋势，在资源短缺的现阶段尤为重要。这几年，我国对绿色建筑由理念到实践，稳步发展，形成了体系的设计方法，各种新技术、新材料也促进了绿色建筑的发展。但必须指出，在建筑设计过程中，不能一味地把国外的高新技术策略生搬硬套到我们的建筑上，对外来技术要有取有舍，应按照中国国情、特定的环境和气候特点来选择适宜的技术，同时利用国产材料降低成本，从可持续角度赋予建筑设计一种新的理念。要实现建筑设计的绿色化，必须引入"整合设计"思想，注意技术集成与优化，最大程度地发挥技术在绿色建筑中的功效。我们应充分发挥创造性和主动性，依靠多专业、多学科的共同合作，在研究系统理论的同时结合工程实践不断探索、验证和总结，创造出具有中国特色的绿色建筑。

东南大学
杨维菊

参考文献：
[1] 仇保兴. 中国绿色建筑发展前景及对策建议[J]. 建筑科技，2011/06.
[2] 李元，秦琴. 从干城章嘉公寓看查尔斯·柯里亚的对空空间[J]. 山西建筑，2008/11.
[3] 刘加平，谭良斌. 浅谈绿色建筑发展[J]. 城市建筑，2008/04.
[4] 王宇哲. 试论绿色建筑理念在建筑设计中的应用.
[5] 程志军. 我国绿色建筑标识项目回顾（2008-2010）[J]. 生态城市与绿色建筑，2011/01.
[6] 张震. 我国绿色建筑的现状和对策[J]. 中国科技信息，2008/16.

CONTENTS

012   **CSET, CENTRE FOR SUSTAINABLE ENERGY TECHNOLOGIES**
可持续能源技术中心

022   **SINO-ITALIAN ECOLOGICAL AND ENERGY EFFICIENT BUILDING (SIEEB), TSINGHUA UNIVERSITY**
清华大学中意环境节能楼

034   **PARKVIEW GREEN**
乔福芳草地

044   **WUXI SUNTECH POWER COMPANY PHASE I**
无锡尚德太阳能电力有限公司一期工程

054   **KUNSHAN NUCLEIC ACID SCIENCE AND TECHNOLOGY PARK**
昆山小核酸产业科技园行政中心

066   **QINGPU ENVIRONMENTAL MONITORING STATION**
上海青浦环境监测站

076   **XI'AN JIAOTONG-LIVERPOOL UNIVERSITY**
西安交通利物浦大学

084   **RENOVATION OF NO.9 XINGHAI STREET, SUZHOU INDUSTRIAL PARK**
苏州工业园星海街9号改造工程

102   **THE GRAND PERGOLA – ADMINISTRATION CENTRE OF BIOLOGICAL OFFICE PARK SUZHOU**
苏州生物纳米科技园

110   **KPMG-CCTF COMMUNITY CENTRE**
毕马威安康社区中心

120   **ZHANGJIAWO ELEMENTARY SCHOOL**
天津西青区张家窝镇小学

134   **SWIMMING HALL AT TAIWAN BUSINESSMEN'S DONGGUAN SCHOOL (TBDS)**
水合院——东莞台商子弟学校游泳馆

目 录

- 146 ZED PAVILION
  上海世博会零碳馆

- 156 NANJING ZIDONG INTERNATIONAL INVESTMENT SERVICE CENTRE OFFICE BUILDING
  南京紫东国际招商中心办公楼

- 166 HUAWEI RESEARCH AND DEVELOPMENT PARK
  华为研发科技园区

- 176 ALIBABA HEADQUARTERS
  阿里巴巴新园区

- 184 SHANGHAI INTERNATIONAL CRUISE TERMINAL
  上海国际港客运中心

- 194 VANKE CENTRE
  万科中心

- 214 LINKED HYBRID
  北京当代万国城

- 226 CHINA NATIONAL OFFSHORE OIL CORPORATION
  中国海洋石油总公司

- 234 BEA FINANCIAL TOWER
  东亚银行金融大厦

- 244 RIVIERA TWINSTAR SQUARE
  浦江双辉大厦

- 256 INDEX
  索引

CSET, CENTRE FOR SUSTAINABLE ENERGY TECHNOLOGIES
Ningbo, Zhejiang Province
Mario Cucinella Architects

可持续能源技术中心

浙江省 宁波市

马里奥·库西尼拉建筑事务所

Gross Floor Area: 1,200m²
Completion Time: 2008
Architect: Mario Cucinella Architects
Photographer: Daniele Domenicali
Award: The Chicago Athenaeum International Architectural Award 2009; MIPIM Green Building Award 2009; SPACE Award 2009

建筑面积：1200平方米

建成时间：2008年

建筑设计：马里奥·库西尼拉建筑事务所

摄影师：丹尼尔·多梅里卡利

奖项：2009年芝加哥建筑设计博物馆国际建筑大奖；
2009年国际房地产交易会绿色建筑奖；2009年空间大奖

The Centre for Sustainable Energy Technologies Building
The new building will provide laboratory, office and seminar accommodation spaces and has been designed to serve as an exemplar building, demonstrating state-of-the-art techniques for environmentally responsible, sustainable construction and energy-efficient internal environmental control. At the same time, it has been designed to minimise its environmental impact by promoting energy efficiency, generating its own energy from renewable sources, and using locally available materials with low embodied energy wherever possible.

Low Carbon Design – Environmental Design Strategy
The CSET building has been designed to respond to diurnal and seasonal variations in ambient conditions by means of a five-point environmental design strategy:
1. High Performance Envelope
2. Exposed Thermal Mass
3. Daylight & Solar Control
4. Natural Ventilation to Tower
5. Piped Ventilation to Laboratory & Workshop

In this way, the building is designed to minimise the need for additional energy for heating, cooling and ventilation. In fact, the residual heating, cooling and ventilation load is estimated to be so low that this residual load, plus demand for electrical power for computing, lighting, etc. will be met from renewable energy sources, including: Ground Source Heat Pump, Solar Absorption Cooling and Photovoltaic Panels.

The spaces within the building have been configured to support a number of different heating, cooling and ventilation strategies, as a demonstration of alternatives to convention heating and cooling systems. Simultaneouslly, renewable and sustainable energy technologies provide the residual heating and cooling requirements, while energy for power and artificial lighting requirements will primarily be met from the large photovoltaic array located to the south of the building. Other renewable energy technologies include solar thermal collectors (linked to a vapour absorption cooling system), a ground-source heat pump (linked to heating/cooling coils within the floor slabs), and wind turbine (for experimental/demonstration purposes).

The building is an outstanding example of contemporary architecture, embodying the aspirations of the researchers and staff who will occupy it, in demonstrating the contribution that sustainable energy technologies can make to the low carbon economy of the future.

Environmental & Energy Performance
The design intention of the building is that it will not require conventional heating and cooling systems and that the residual energy requirement will be met by renewable energy sources, thus minimising its carbon footprint. It has also been designed to respond to the diurnal and seasonal variation in the climate of Ningbo, to minimise heating requirement in winter and cooling in summer, and to promote natural ventilation in spring and autumn when environmental conditions allow. The building is therefore well insulated, incorporates high thermal capacitance internal floors and walls, and a ventilated glazed south façade. In the cold period, the only additional heat required will be to pre-heat ventilation air, and (when it is very cold outside) to raise internal surface temperatures. To this end the south façade helps to passively pre-heat ventilation air supplied by natural convection to teaching rooms, offices and meeting rooms. Air supplied (by fans) to the workshop and laboratory is pre-heated via tubes in the ground. A reversible ground source heat pump will also be utilised to provide "top-up" heating through coils embedded in the soffit of the concrete floors.

1. Façade detail
2. Front view
3. General view
1. 外墙特写
2. 正面特写
3. 全景

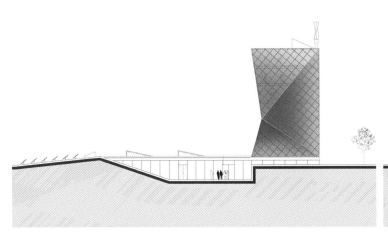
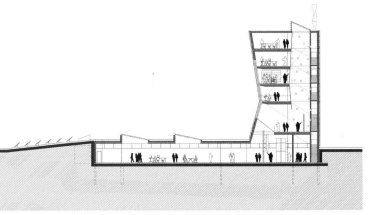

Cooling

In summer, the high performance envelope and the thermal capacitance of the exposed concrete surfaces internally, will generally keep the interior cool. The only additional cooling required will be to pre-cool the ventilation air and (when very hot outside) to reduce the surface temperatures. To this end, air supplied to the workshop and laboratory is passively pre-cooled via the ground tubes and then dehumidified and cooled by an air handling unit located in the basement. Air supplied to the tower is dehumidified and mechanically cooled by an air handling unit located at the roof top, then introduced to the top of the lightwell, falling down to each level, from which it is exhausted by the naturally ventilated façade. The solar collectors would provide the absorption package chiller with the required energy to deliver cooling to the two air handling units. In addition the reversible ground source heat pump will provide cooling to the ceiling of the concrete floors.

Ventilation

During the mid-seasons (spring and autumn), natural ventilation is promoted in most spaces, controlled automatically by means of vent opening gear within the perimeter glazing. During the summer, when it is both hot and humid, it is necessary to de-humidify and cool the supply air, and the electrical power for this is provided by the photovoltaic system.

Lighting

The building has been designed to exploit day lighting as far as possible, while avoiding glare and solar heat gain. This reduces the amount of time for which artificial lighting is required.

The Photovoltaic

(PV) Solar system will be used to provide artificial lighting and small power for office equipment such as computers and fax machines. During the peak period of sunshine enough power shall be produced from the PV system to run other equipment such as the lift, the mechanical ventilation and chilled water systems. In the event of extra power not being utilised, it shall be stored in batteries or transferred to the nearby sport centre.

Building Management System

The building is equipped with a management system dedicated to the electrical and mechanical plants for optimising electrical loads and reducing energy consumption. The system will allow centralisation of controls and signals of the building's technological equipments. The installed software will allow commands to be sent automatically to all field actuators and equipments.

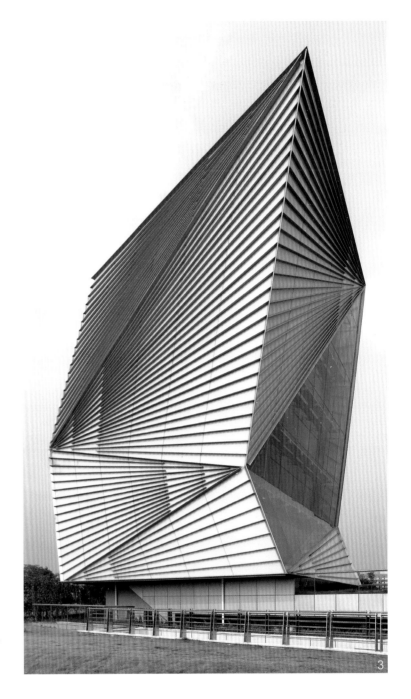

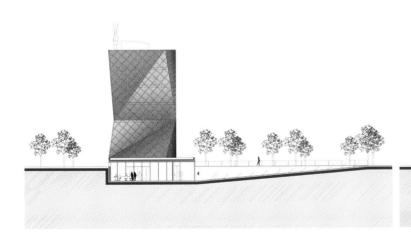

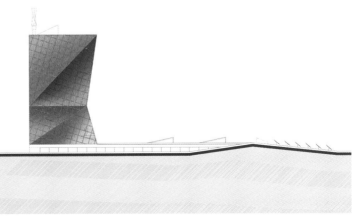

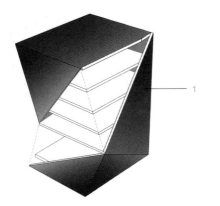 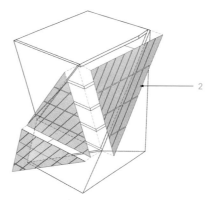 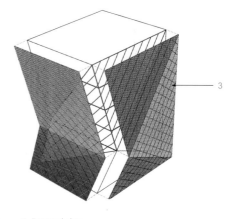

1. Insulated concrete wall
   Internal concrete structure wall
   with external insulation panels
   -thermal mass
   -high thermal insulation
   1. 隔热混凝土墙
   内部混凝土结构墙和外部保温板
   – 蓄热体
   – 高度保温层

2. Structural double skin façade (U-value 1.4-1.3w/mgk)
   Structural façade with thermally
   broken painted aluminium profiles
   with double glazing units (8/20/6mm)
   -Natural light
   -High thermal insulation
   2. 结构双层立面（U-value 1.4-1.3w/mgk）
   结构立面，配有热破坏喷漆铝截面及
   双层玻璃单元（8/20/6毫米）
   – 自然采光
   – 高度保温层

3. External skin
   External façade with laminated glass and silk
   screened pattern on inner pane (5+5+0.76mm)
   External façade elevation detail
   -Solar protection
   -Natural light
   -Natural ventilation
   3. 外壳
   外立面配有夹层玻璃，
   内层窗格上是丝印图案（5+5+0.76毫米）
   外立面细部
   – 日光保护
   – 自然采光
   – 自然通风

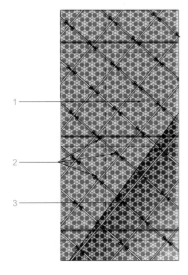 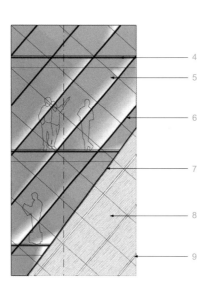 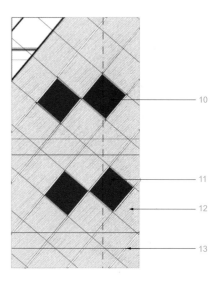

External Façade Elevation Detail (Above)
1. External façade
   With laminated glass and silk screened
   pattern in inner pane (5+5+0.78mm)
2. Points for mechanical fixing of the
   external façade to the internal façade
3. Points for mechanical fixing
   of the external façade to
   the concrete structure

外立面细部：
1. 外立面
   配有夹层玻璃，
   内层窗格上是丝印图案（5+5+0.78毫米）
2. 外立面以机械固定在内立面上
3. 外立面以机械固定在混凝土结构上

Internal Façade Elevation Detail (Above)
4. Horizontal transom corresponding to the slab
5. Internal structural façade
   with thermally broken painted
   aluminium profiles divided
   in tilted fixed double glazing units
   (8/20/4+4.2mm)
6. Tilted mullion
7. Tilted joint line between concrete
   wall and internal glass façade
8. Insulation panel
   With aluminium finishing 50mm thk
   Panels are mechanically fixed
   to the concrete structure
9. Aluminium joint between
   the insulation panels

内立面细部：
4. 与平板一致的水平气窗
5. 内结构立面
   配有由倾斜双层玻璃单元分割的热破坏
   喷漆铝截面（8/20/4+4.2毫米）
6. 倾斜窗框
7. 混凝土墙和内层玻璃的倾斜接合线
8. 保温板
   配有50毫米厚的铝装饰层
   面板以机械固定在混凝土结构上
9. 保温板之间的铝接缝

Internal Façade Elevation Detail – Concrete Façade (Above)
10. Internal openable insulated panel for natural ventilation
    -Openings are inserted into the façade grid pattern
    -Number of openings to be decided
    according to environmental strategies
11. Internal openable panel for natural ventilation
12. Insulation panel
    with aluminium finishing 50mm thk
    Panels are mechanically fixed to the concrete structure
13. Aluminium joint between the insulation panels

内立面细部（混凝土层）：
10. 内部可开式保温板
    保证自然通风
    开口嵌入外立面的栅格图案
    开口的数量由环境策略决定
11. 内部可控式面板
    保证通风
12. 保温板
    配有50毫米厚的铝装饰层
    面板以机械固定在混凝土结构上
13. 保温板之间的铝接缝

4. View above entrance
4. 入口上方

5. Entrance view
5. 入口

**Energy Strategies Section-Summer/Winter
(21 June_12am / 21 Dec_12am)**
1. Solar
   A chiller, power by hot water from solar tubes, pre-cool external air for ventilation of the tower
2. Exhaust air
3. Closed
4. Light well
5. High thermal inertia of the exposed concrete surfaces
6. Double skin façade solar and glare control
7. Thermal mass activation
8. Radiant cooling
9. Green roof
   High thermal inertia avoids overheating of the interior
10. External air
11. Green spaces reduce the heat island effect
12. Earth-to-air heat exchanger
13. Underground pipes pre-cool air for the semi-basement
14. Bypass closed
15. Cooling dehumidifying coil
16. A bms manage active and passive strategies to minimise energy consumption
17. No.16 vertical geothermal loops
18. Electricity from pv
19. Hot water from solar collectors_114m^2 evacuated cubes
20. Reversible-cycle heat pump
21. In the sunny days the double skin façade pre-heats ventilation air
22. Underground pipes pre-heat air for the semi-basement
23. Bypass open
24. Well insulated and air tight envelope
    Opaque wall u=0.25 W/mqk
    Transparent façade u=1.2 W/mqk
    U=0.25 W/mqk
    U=1.2 W/mqk

能源策略剖面图–夏季/冬季
（6月21日 中午12点 / 12月21日 中午12点）
1. 日光
   冷却装置，由太阳能管内的热水提供电力，预先冷却塔楼内通风的外部空气
2. 排气
3. 闭合
4. 光井
5. 露石混凝土表面的高热惯性
6. 双层立面日光和强光控制
7. 热质量激活
8. 辐射制冷
9. 绿色屋顶
   高热惯性避免室内过热
10. 外部空气
11. 绿色空间减少了热岛效应
12. 地对空的热交换
13. 地下管道为半地下室的空气预制冷
14. 支路闭合
15. 冷却除湿线圈
16. 电池管理系统采用主动和被动策略来减少能源消耗
17. 16号垂直地热循环
18. 来自光伏电池的电力
19. 太阳能热水器提供的热水 114平方米排空体
20. 可逆循环热泵
21. 阳光充足时，双层立面能够为流通空气预热
22. 地下管道为半地下室预热空气
23. 支路开放
24. 良好的保温层和密封外壳
    封闭墙U=0.25 W/mqk
    透明立面U=1.2 W/mqk

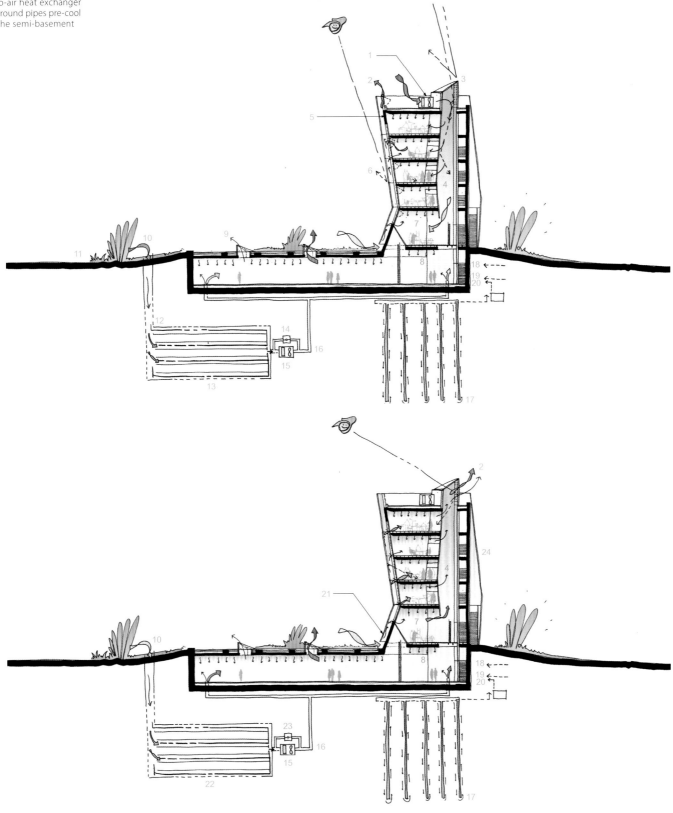

可持续能源技术研究中心

可持续能源技术研究中心提供实验室、办公室和研究室，它也是一座模范建筑，展示了最新的环保、可持续建造和节能内部环境控制技术。建筑通过提升能源效率、从可再生能源中获取能量和使用当地的低能耗材料，将自身对环境的影响降到了最低。

低碳设计——环境设计策略

可持续能源技术研究中心通过五个角度的环境设计策略反映了每日、每季外界条件的不同，这包括：高性能外壳、外露蓄热体、日光和阳光控制、高楼自然通风以及实验室和工作室管道通风。

这样一来，建筑就能够将用于供暖、制冷和通风的额外能源需求量降到最低。事实上，余下的供暖、制冷和换气负荷极低，与计算机、照明等的电力需求加在一起，通过可再生能源都能得到满足。这些可再生能源包括：地源热泵、太阳能吸收式制冷和太阳能光电板。

建筑内部经过配置，支撑了一系列不同供暖、制冷和通风策略，替代了传统的供暖和制冷系统。可再生和可持续能源策略满足了余下的功能和制冷需求；位于建筑南面的大型太阳能光电池组则提供电器和照明所需的能源。其他策略包括太阳能集热器（与蒸汽吸收制冷系统相连）、地源热泵（与楼板内的供热/制冷盘管相连）和风涡轮（用于实验和展示）。

研究中心是现代建筑的杰出代表，体现了中心研究人员和员工的愿望，展示了可持续能源技术能够为未来低碳经济作出的贡献。

环境和能源性能

建筑的设计意图是：无需传统供暖和制冷系统，让可再生能源满足剩余的能源需求，从而将碳排放量最小化。设计还反映了宁波每日、每季的变化，尽量减少供暖或夏季需求，并促进自然通风。因此，建筑隔热良好，结合高热容室内楼板和墙壁，并在南外立面引入通风玻璃外墙。在寒冷期，唯一的附加热需求来自预热通风空气和提升内表面温度（外面极冷时）。此时，南立面将通过教室、办公室和会议室之间的自然对流帮助室内被动地预热空气。工作室和实验室的空气供给（通过电扇）通过地下涡轮预热。可逆式地源热泵将通过混凝土地面下的盘管提供"追加"供暖。

制冷

在夏季，建筑的高性能外壳和内部外露混凝土表面的热容将保证室内的凉爽。唯一的附加制冷需求来自预冷却通风空气和降低建筑内表面温度（当外面极热时）。此时，工作室和实验室通过地下管道进行被动式预制冷，然后通过地下室的空气处理装置进行干燥和冷却工作。高楼供给的空气通过屋顶的空气处理装置进行干燥和冷却，然后引到采光井的顶部，逐次降到各个楼层，通过自然通风的外立面耗尽。太阳能集热器将为吸收式冷却器提供能源，从而将制冷传递到两个空气处理装置。此外，可逆式地源热泵将为混凝土楼面的天花板提供制冷。

通风

春秋两季，大多数空间都通过自动控制玻璃通风口装置来提升自然通风条件。在湿热的夏季，太阳能光电板提供干燥和冷却空气所需的电力。

照明

建筑的设计最大限度地利用了日光照明，同时又避免了强光和过度的太阳辐射量。这一设计减少了人工照明的时间。

太阳能光电板

太阳能光电板系统将提供人工照明和电脑、传真机等办公设备所需电量。在日照高峰期，光电板系统能为电梯、机械通风和冷却水系统提供电力。当能源过量时，它们将被存储在电池里或转移到附近的运动中心。

建筑管理系统

建筑配有一个专门的电力和机械管理系统，以优化电力负荷、减少能源消耗。该系统能够集中控制建筑的技术设备。系统安装的软件能够自动发出指令，控制所有致动器和设备。

6. Interior view from first floor to basement
7, 8. Interior view
6. 一楼至地下室
7、8. 室内

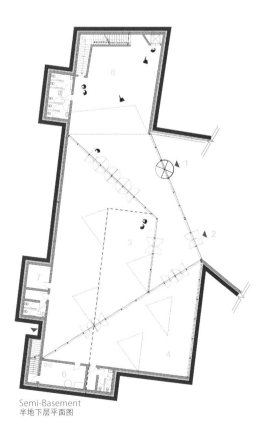

Semi-Basement
半地下层平面图

Floor Plans:
1. Main entrance
2. Service entrance
3. Laboratory
4. Workshop
5. WC
6. Plant room
7. Storage
8. Expo space/reception
9. Expo store
10. Office
11. Teaching room
12. Meeting room
13. Kitchenette
14. Roof garden
15. Expo space

楼层平面图:
1. 主入口
2. 服务区入口
3. 实验室
4. 车间
5. 卫生间
6. 机房
7. 储藏室
8. EXPO 空间/接待室
9. EXPO 零售店
10. 办公室
11. 教室
12. 会议室
13. 小厨房
14. 屋顶花园
15. EXPO 空间

First Floor
一层平面图

Second Floor
二层平面图

Third Floor
三层平面图

Fouth Floor
四层平面图

Fifth Floor
五层平面图

First Floor
一层平面图

SINO-ITALIAN ECOLOGICAL AND ENERGY EFFICIENT BUILDING (SIEEB), TSINGHUA UNIVERSITY
Beijing
Mario Cucinella Architects

清华大学中意环境节能楼
北京
马里奥·库西尼拉建筑事务所

Gross Floor Area: 20,000m²
Completion Time: 2006
Architect: Mario Cucinella Architects
Photographer: Daniele Domenicali, Alessandro Digaetano, MCA Archives
Award: 2007 Chicago Anthenaeum Museum of Architecture and Design – International Architecture Award

建筑面积：20000平方米
建成时间：2006年
建筑设计：马里奥·库西尼拉建筑事务所
摄影师：丹尼尔·多蒙尼卡利，亚历山德罗·迪盖塔诺，马里奥·库西尼拉建筑事务所
奖项：2007年芝加哥雅典娜神庙奖——国际建筑大奖

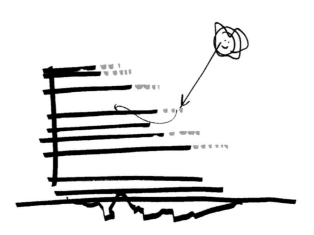

This design philosophy combines sustainable design principles and state-of-the-art technologies to create a building that responds to its climatic and architectural context. The design uses both active and passive strategies through the design of its shape and of its envelope to control the external environment in order to optimise the internal environmental comfort conditions.

Sino-Italian Ecological and Energy Efficient Building was presented on December 7th 2004 in Beijing during the visit of Italian President Ciampi. The SIEEB project is the ideal result of cooperation between the Ministry for Environment and Territory of the Republic of Italy and the Ministry of Science and Technology of the People's Republic of China and is also regarded as a platform to develop the bilateral long-term cooperation in the environment and energy fields and a model case for showing the CO_2 emission reduction potential in the building sector in China.

This building is realised in the Tsinghua University Campus in Beijing and has been designed by architects Mario Cucinella and Politecnico of Milan. It is a 20,000-square-metre building, forty metres high and it will host a Sino-Italy education, training and research centre for environment protection and energy conservation. This SIEEB project is the result of a collaborative experience among consultants, researchers and architects. This integrated design process is a most distinctive part of the project and a key issue for green buildings. The building is therefore generated through a series of testing and computer simulations of its performance in relation to its possible shape, orientation, envelope, technological systems and so on. The building is designed to find a balance among energy efficiency targets, minimum CO_2 emissions, a functional layout and the image of a contemporary building.

The envelope components, as well as the control systems and the other technologies are the expression of the most updated Italian production,

1. Façade
2. General view
3. East view
1. 外立面
2. 全景
3. 东立面

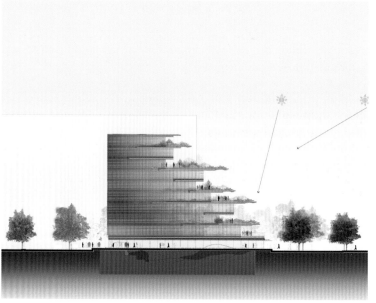

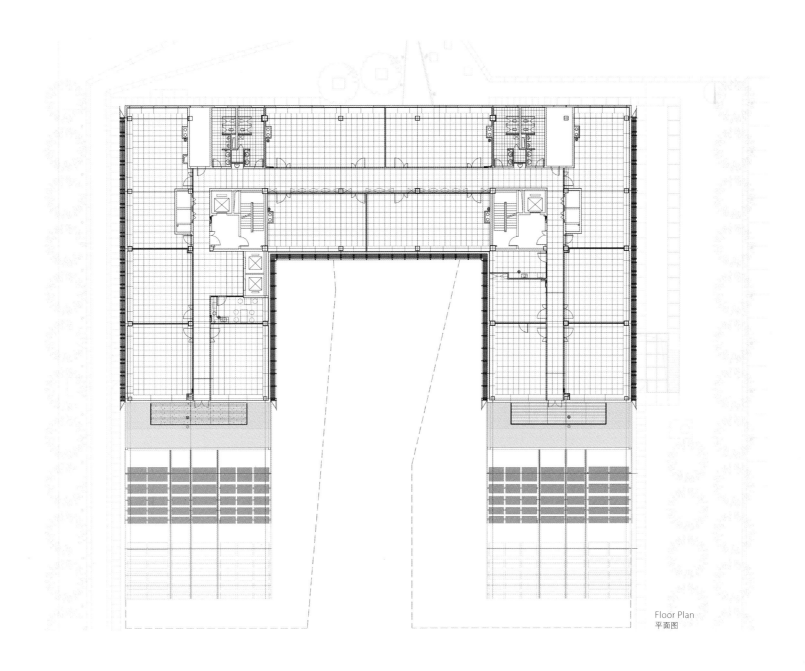

Floor Plan
平面图

within the framework of a design philosophy in which proven components are integrated in innovative systems.

The SIEEB building shape derives from the analysis of the site and of the specific climatic conditions of Beijing.

Located in a dense urban context, surrounded by some high-rise buildings, the building optimises the need for solar energy in winter and for solar protection in summer.

Reflecting and semi-reflecting lamellas and louvres will also allow for sunshine to penetrate in the rooms in winter and to be rejected in summer, reducing the energy consumption of the building.

Artificial lighting will be based on high efficiency lamps and fittings, controlled by a dimming system capable to adjust the lamps power to the actual local lighting needs, combinating with the natural light contribution. A presence control system will switch off lights in empty rooms.

Thermal comfort conditions are provided by a primary air (distributed by means of a displacement ventilation system) + radiant ceiling system. This combination minimises electricity consumption in pumps and fans.

Lightweight radiant ceilings allow for lower air temperature in winter and higher in summer, thus reducing energy consumption; moreover, the presence sensors, coupled with CO_2 sensors, can modulate either the air flow or the ceiling temperature when few or no people are in the room, thus avoiding useless energy consumption. In summer night cooling takes place.

Gas engines are the core of the energy system of the building. They are coupled to electric generators to produce most of the electricity required. The engines waste heat is used for heating in winter, for cooling – by means of absorption chillers – in summer and for hot water production all year round.

A sophisticated, "intelligent" control system manages the plant. Because of the cleaner electricity produced the amount of CO_2 emissions per square metre of the SIEEB will be far lower than in present Chinese commercial building stock.

4. Environment
5. View from ground
4. 环境
5. 仰拍

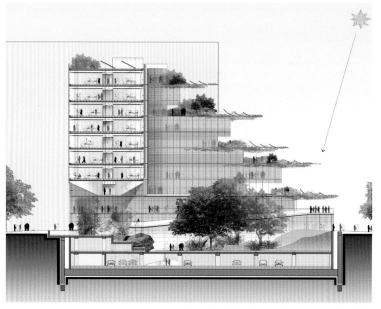

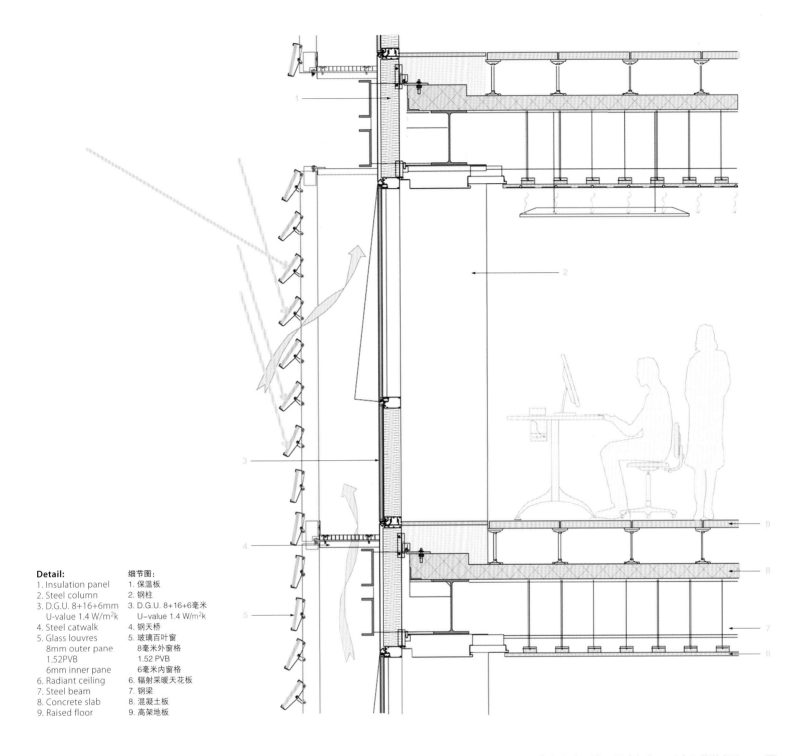

Detail:
1. Insulation panel
2. Steel column
3. D.G.U. 8+16+6mm
   U-value 1.4 W/m^{2}k
4. Steel catwalk
5. Glass louvres
   8mm outer pane
   1.52PVB
   6mm inner pane
6. Radiant ceiling
7. Steel beam
8. Concrete slab
9. Raised floor

细节图：
1. 保温板
2. 钢柱
3. D.G.U. 8+16+6毫米
   U-value 1.4 W/m^{2}k
4. 钢天桥
5. 玻璃百叶窗
   8毫米外窗格
   1.52 PVB
   6毫米内窗格
6. 辐射采暖天花板
7. 钢梁
8. 混凝土板
9. 高架地板

　　清华大学的中意生态节能楼的设计理念融合了可持续原则与建筑技术的艺术，创造了一座与当地气候环境和建筑风格相适应的建筑。在对其外形的设计中，通过控制外部环境，使内部环境的舒适度最大化。

　　该项目于2004年12月7日意大利总统钱皮访华之际在北京揭幕，是意大利环境与领土司与中国科技部合作的产物，被视为双方在环境和能源领域长期合作的平台，体现了中国建筑领域如何实施二氧化碳减排的标准。

　　大楼位于清华大学校园内，由建筑师马里奥·库西尼拉和米兰理工大学共同设计。大楼占地20000平方米，高40米，作为中意环境保护和能源保护的教育、培训和研究中心。集合了专业人士的智慧是该项目最独特之处，并且经过一系列对建筑各方面表现的测试和电脑运算，包括可能的外形、朝向和薄膜技术系统等。目的在于在能源高效利用、二氧化碳最低排放、功能性布局和当代建筑的形象之间寻找一个平衡点。

　　建筑表面元素、控制系统和其他技术都代表了意大利最先进的生产水平，整个设计理念的框架就是：将可靠的组成部分融合在创新的系统中。

　　清华大学环境节能楼的形态设计是经过对当地的仔细分析，并且考虑到北京特殊的气候条件之后而决定的。大楼处于人口密集的城市地带，周围有很多高层建筑，大楼的这种设计既要确保冬天对太阳能的摄取，又要确保夏天日照不会过于强烈。全反射和半反射的薄板和百叶窗使得阳光冬天可以照射进房间，夏天则被拒之窗外，降低了整幢大楼的能源消耗。

　　人工照明采用了高效的电灯和电器设备，用一个调光系统控制，能调节电灯的亮度使其与自然光线相适应，满足当地实际的照明需要。感应控制系统会自动关掉无人使用的房间里的灯。

　　温度的舒适是由一个基本的空气（来自排通风系统）和辐射性天花板系统实现的。这样的组合让泵和风扇用电量降到最低。

　　轻质的辐射采暖天花板使得冬天里空气的温度降低而夏天则会升高，这样就减少了能源的消耗。另外，感知感应器和二氧化碳感应器共同作用，可以在房间里人很少或没有人的时候调节空气的流动和天花板的温度，避免了能源浪费。夏天，夜晚的降温系统将开始工作。

　　天然气引擎是能源系统的核心，与电动发电机共同作用，提供大部分电力。引擎产生的废热冬天用于供暖，夏天通过吸收冷却器降温，还可供应全年热水。一个智能控制系统控制着这一系统。凭借更清洁的电力，这幢节能楼每平方米的二氧化碳排放量远远低于现在中国的商业大楼。

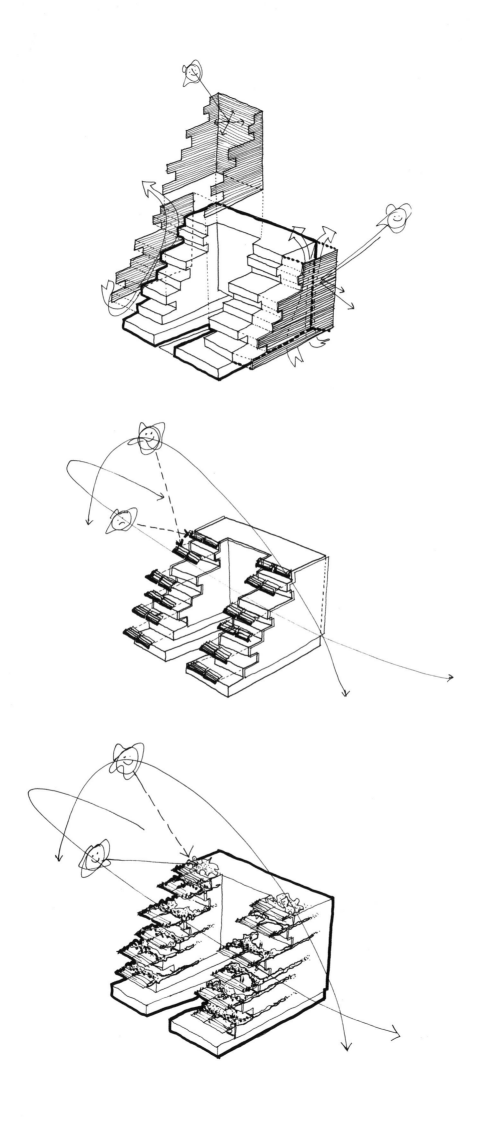

6. Terraces
6. 露台

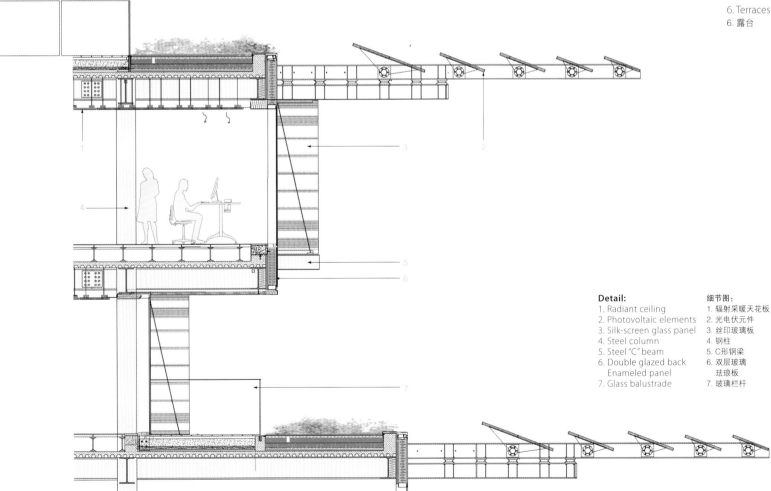

Detail:
1. Radiant ceiling
2. Photovoltaic elements
3. Silk-screen glass panel
4. Steel column
5. Steel "C" beam
6. Double glazed back Enameled panel
7. Glass balustrade

细节图：
1. 辐射采暖天花板
2. 光电伏元件
3. 丝印玻璃板
4. 钢柱
5. C形钢梁
6. 双层玻璃珐琅板
7. 玻璃栏杆

7. Northwest view
8. Northwest corner
7. 西北立面
8. 西北角

Details (Right Two):
1. Insulation panel
2. D.G.U. 8+16+6mm
   U-value 1.4 W/m^{2}k
3. Radiant ceiling
4. Steel beam
5. Concrete slab
6. Insulation panel
7. Raised floor
8. Clear float panes
   with horizontal silkscreens
   10mm outer panes
   1.52 PVB
   6mm inner panes
9. Internal aluminium light-shelf

细节图（右方两图）：
1. 保温板
2. D.G.U. 8+16+6毫米
   U-value 1.4 W/m^{2}k
3. 辐射采暖天花板
4. 钢梁
5. 混凝土板
6. 保温板
7. 高架地板
8. 配有丝印的清除浮动窗格
   10毫米外窗格
   1.52 PVB
   6毫米内窗格
9. 内部铝制光架

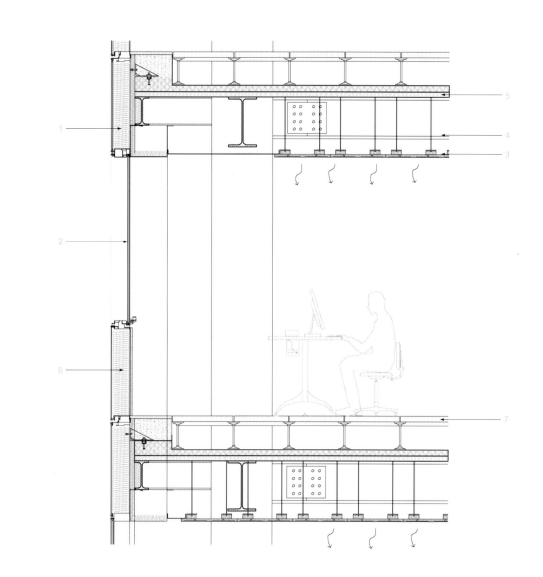

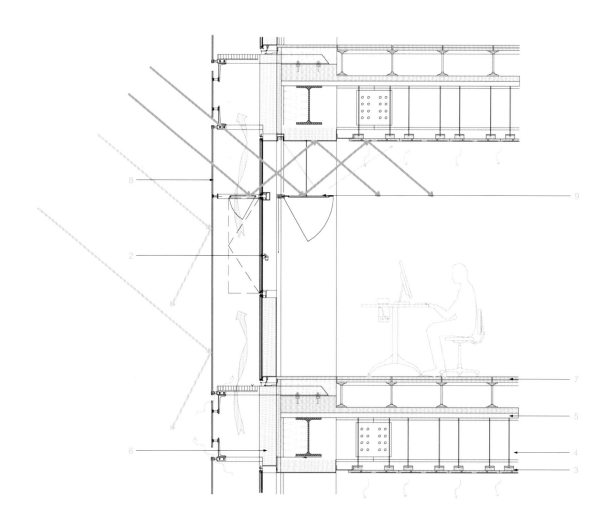

031

9. Courtyard
9. 庭院

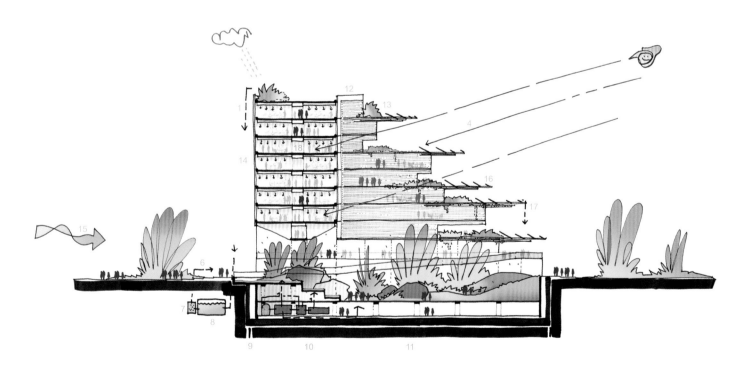

Details (Above Two):
1. Rainwater collection
2. Passive solar design provides shading in summer
3. Radiant cooling
4. Photovoltaic panels placed above the south façade provide an overhanging surface which shades the glazed wall from sun radiation
5. Green spaces and a water pond reduce the heat island effect and contribute to passive cooling
6. Irrigation
7. Water recovery unit
8. Rainwater tank
9. Cogenerator
10. Absorption heat pump
11. A bms manages active and passive strategies to minimise energy consumption
12. Double skin façade U=1.4W/m²k
13. Passive solar design provides solar gains in winter
14. North façade U=1.4W/m²k FS=43%
15. Planting protects the building from cold winter winds
16. Deciduous planting allows solar gains and day lighting in winter
17. Power electricity
18. Radiant heating

细节图（上方两图）：
1. 雨水收集
2. 被动式太阳能设计为夏天提供阴凉
3. 辐射制冷
4. 南立面的光电伏板形成悬臂表面，为玻璃幕墙遮挡太阳辐射
5. 绿色空间和水塘减少了热岛效应，促进了被动制冷
6. 灌溉
7. 水回收单元
8. 雨水槽
9. 联合发电机
10. 吸收式热泵
11. 电池管理系统管理主动和被动策略，以最小化能源消耗
12. 双层外立面 U=1.4W/m²k
13. 被动式太阳能设计在冬天提供阳光吸收
14. 北立面 U=1.4W/m²k, FS=43%
15. 植物保护建筑不受冬日寒风侵袭
16. 落叶植物保证了冬天的太阳热量和自然采光
17. 电力
18. 辐射供暖

033

PARKVIEW GREEN
Beijing
Integrated Design Associates Ltd.

乔福芳草地
北京
综合设计事务所

Gross Floor Area: 200,000m²
Completion Time: 2007
Architect: Integrated Design Associates Ltd.
Photographer: Integrated Design Associates Ltd.
Award: Best Green Building in Asia, Mipin Asia 2010

建筑面积：200000平方米
建成时间：2007年
建筑设计：综合设计事务所
摄影师：综合设计事务所
奖项：2010年米品亚洲大奖之亚洲最佳绿色建筑

The design concept is centred around a clear intention to provide the users of the development, comprising of a retail mall, commercial office space, and a six-star boutique hotel, with an internal environment that is tempered by nature to create microclimates that would give comfort to users even in the extreme climates of Beijing, and at the same time reduce the overall energy consumvtion of the building.

The development is made up of four buildings sited in a sunken garden to the surrounding street level. All buildings are designed with atria spaces, sky-gardens, terraces, and link bridges, and together are shielded from the external environment by an outer building envelope that is constructed of steel, glass and ETFE cushions. The skin is essentially the weather protection layer that controls the microclimate of the entire development by the way of a thermal insulation layer which is formed in the airspace that is between the skin of the internal buildings and the outer skin.

The retail mall comprises of around 55,000 square metres in area and occupies the lower levels of the development. The mall wraps around a natural light-filled atria, the lowest level of which is landscaped to provide pockets of tranquil spaces for resting and socialising. The design of the mall is intended to give a unique experience to visitors and this is partly found in the presence of a dramatic large-span suspended footbridge that connects together the two furthest corners of the retail mall, connecting them with the four levels of the retail mall. There are also small number of retail "pods" which are constructed of glass and steel. The "pods" are cantilevered into the atria to provide key locations for "boutique-style" merchandisers to create suspense as part of the overall retail experience.

Climatic Characteristics of Its Location
The summer climate is usually hot and wet with temperatures reaching up to 40 degrees Celsius. Most of the annual precipitation occurs in the summer months. Winter climate is cold and dry with temperatures as low as -20 degrees Celsius. Both spring and autumn temperatures are moderate with temperatures between 15 and 23 degrees Celsius.

Main Eco Features: Hybrid Ventilation and Night Cooling in Office
1. Chilled ceiling and underfloor air-conditioning in offices
2. Double façade creating thermal break effect
3. Sky-gardens on every floor
4. Operable ETFE roof for ventilation and stack effect
5. Residual air-conditioning in sky gardens when required
6. Binnacles providing displacement ventilation at atrium
7. Retail air-conditioned air-reuse for common area
8. Evaporative type water cooled air-conditioning
9. Earth cooling for fresh air pre-cool/pre-heat
10. Demand control ventilation
11. Reuse of basement ventilation air for cooling tower
12. Heat pipe for dehumidification
13. Variable speed pump and ventilation fans
14. Daylight harvesting
15. Rainwater recycling and Greywater recycling
16. Energy efficient luminaries lower overall power outputs (Watts/sqm)
17. Environmental materials and finishes used for interior fitting-out
18. Landscape design using only native plants and trees species

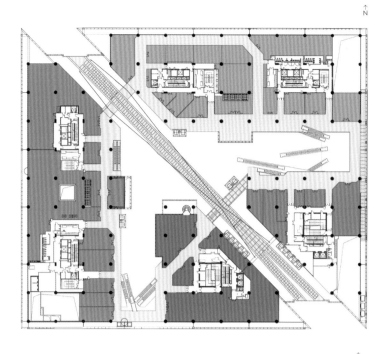
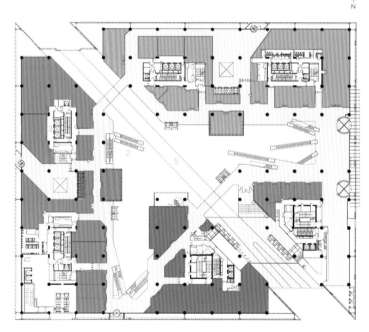

L3 Retail Floor Plan: 零售区三层平面图：
1. Entrance　　1. 入口
2. Lift　　　　 2. 电梯
3. Shopping mall　3. 商场
4. Staircase　　4. 楼梯
5. Footbridge　 5. 人行桥

1. Façade
2. Night view of façade
1. 外观
2. 外观夜景

037

3. Roof to collect water
3. 收集雨水的屋顶

项目的设计理念以明确的意图为中心，即为使用者提供所需的开发工程——购物中心、商务办公空间和一家六星级精品酒店。建筑内部环境要与自然调和，为使用者在北京极端气候中提供舒适的微环境，同时减少建筑总能耗。

项目由四座建筑围绕着一个下沉花园组成，分别设有天井、空中花园、露台和连接桥，四者由一个巨大建筑外壳（由钢材、玻璃和四氟乙烯衬垫组成）罩起来，与外部环境隔绝开来。作为气候保护层，外壳通过由内部建筑和外壳之间的空气隔热层控制着整个项目的微环境。

购物中心总面积55000平方米，占据项目的下层空间，它环绕自然采光的天井而建，最底层为景观设计，提供休息和社交的独立空间；购物中心的设计旨在为访客提供独特的体验。超大跨度的人行桥连接购物中心的两个圆角，使其与购物中心的四层楼相连。由玻璃和钢材建成的小型零售空间悬入天井之中，为精品店提供了空间，为整体购物体验提供了悬念。

项目所在地的气候特征：
夏季炎热潮湿，气温可达40摄氏度。一年的大多数降水都发生在夏季。冬季寒冷干燥，最低气温低至零下20摄氏度。春秋两季气候温和，平均气温在15到23摄氏度之间。

主要生态特征：
1. 办公室混合通风和夜间制冷；
2. 办公室冷天花板和地板下空调系统；
3. 双层外墙打造断热效应；
4. 每层楼都有空中花园；
5. 可控制式四氟乙烯屋顶，形成了通风和烟囱效应；
6. 空中花园的附加空调系统；
7. 沟渠系统为中庭提供了置换通风；
8. 公共区域空气再利用；
9. 水蒸气冷却空调系统；
10. 为新鲜空气预制冷/预热的地面制冷；
11. 需求控制通风；
12. 再利用地下室通风空气来冷却大楼；
13. 热管去湿；
14. 变速泵和换气扇；
15. 日光收集；
16. 雨水和灰水回收利用；
17. 节能照明设施降低了总功率（瓦/平方米）；
18. 室内装修采用环保材料和装饰；
19. 景观设计仅采用本土植物和树种。

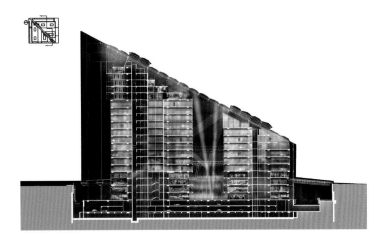

1. Exhaust air discharge
2. Double skin façade
3. Greywater recycling
4. Eco-well for hybrid ventilation
5. Day lighting
6. ETFE roof
7. Landscape sunken garden
8. Greenery
9. Fresh air intake
10. Earth cooling tunnel for fresh air pre-cool/pre-heat
11. Retail AC air reuse for common area
12. Binnacle for displacement ventilation at atrium
13. Demand control ventilation
14. Basement ventilation air reuse for cooling tower
15. Water-cooled air conditioning system
16. Chilled ceiling & underfloor AC
17. Operable window for office natural ventilation
18. Underfloor AC

1. 废气排放
2. 双层外立面
3. 灰水循环
4. 生态井，保证综合通风
5. 自然采光
6. 四氟乙烯屋顶
7. 景观下沉花园
8. 绿化
9. 新鲜空气进气
10. 地下制冷管为新鲜空气预制冷/预热
11. 交流通风被再利用于公共区域
12. 中庭置换通风的罗盘箱
13. 需求控制通风
14. 地下通风被再利用于制冷塔
15. 水制冷空调系统
16. 冷却天花板和地下交流通风
17. 为办公室提供自然通风的可控式窗
18. 地下交流通风

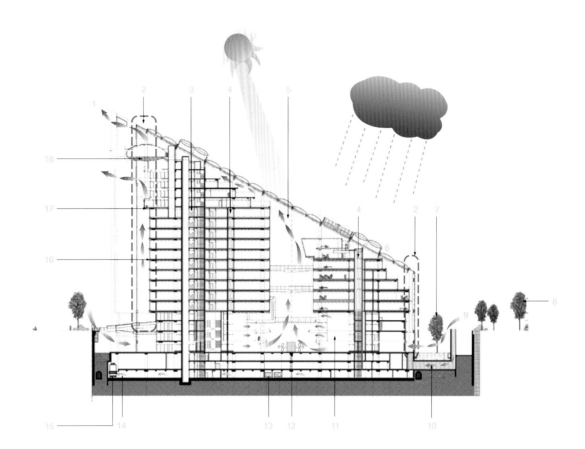

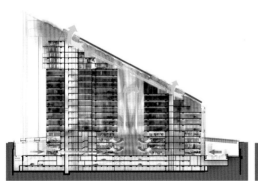

Spring & Autumn Seasons
春季和秋季

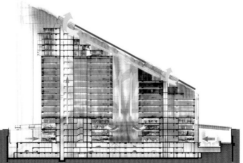

Summer Season
夏季

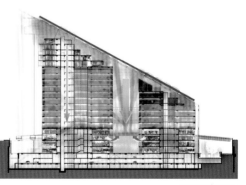

Winter Season
冬季

039

4. Office towers
5. View of internal suspension bridge from open public space of the mall
6. 236-metre-long internal suspension bridge
4. 塔楼
5. 在购物中心广场上仰望钢架步行桥
6. 长236米的钢架步行桥

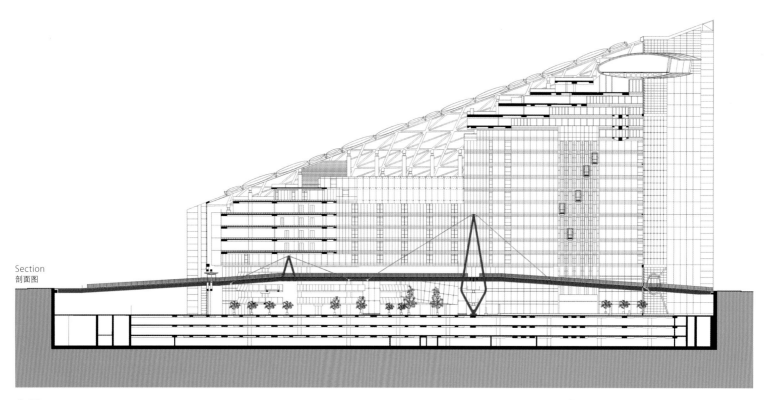

Section
剖面图

7. Corridor of HOTEL ÉCLAT BEIJING
8. HOTEL ÉCLAT BEIJING
9. Office space
7. 北京怡亨精品酒店的走廊
8. 北京怡亨精品酒店
9. 办公空间

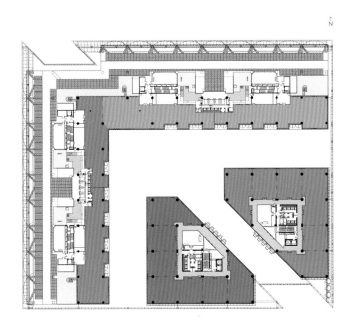
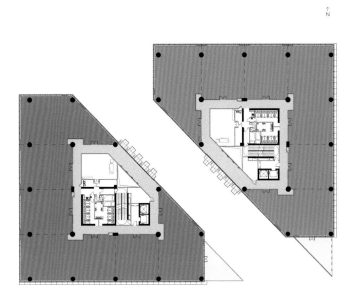

WUXI SUNTECH POWER COMPANY PHASE I
Wuxi, Jiangsu Province
Suzhou Institute of Architectural Design Co., Ltd.

无锡尚德太阳能电力有限公司一期工程

江苏省 无锡市

苏州设计研究院股份有限公司

Gross Floor Area: 54,000m²
Completion Time: 2009
Architect: Suzhou Institute of Architectural Design Co., Ltd.
Design Team: JIN Jianhua, WANG Zhiyong, HUANG Peng, JIANG Hua
Photographer: ZHA Zhengfeng
Award: 2011 First Prize of Jiangsu Excellenct Architecture Design

建筑面积：54000平方米
建成时间：2009年
建筑设计：苏州设计研究院股份有限公司
设计团队：靳建华，王智勇，黄鹏，蒋婳
摄影师：查正风
奖项：江苏省优秀建筑设计一等奖

Overview
The central issue of this project is combining architecture form and energy conservation technology, excavating the protential of renewable energy in creation, enhancing renewable energy's application in architecture and achieving architectural enenry generation. The minimalist appearance of the research and development building and its rich and inspiring interior space make a strong contrast, which is the most noticeable spot.

The site is divided into manufacture area and office area. The circulation system is created according to transportation requirement and fire protection arrangement. The whole architecture functionally consists of three parts: manufacture building, office/R&D building and recreation centre. The designers try their best to create a modern cooperate identity of novel appearance and beautiful environment.

Architecture Form
Based on the complicated function and location, the office/R&D building and recreation centre become the main parts of the architecture. The R&D building is divided into seven levels according to their functions. Each level is further divided into different areas, forming a whole system through the public path adjacent to manufecture building. At the same level, the different areas intertwine with each other, creating a rich sense of space. The recreation centre makes the whole architecture into a unity. The court, gym and restaurant are independently located in the centre of the space. The different function areas communicate and blend together through the same space.

The four-storey manufacture building is 54,000 square metres. The main structure is made of reinforced concrete. The design of the manufacture building takes economy and functionality as its principle. The simple and pure identity will present the product's high technology.

Site Plan (Below):
1. Existing plant
2. New R&D building
3. Office building
4. Recreation centre
5. Storage for dangerous goods and special material gas
6. Substation
7. Parking

总平面图（下图）：
1. 已有厂房
2. 新增生产及研发楼
3. 办公楼
4. 康乐中心
5. 危险品库和特气库
6. 变电站
7. 停车场

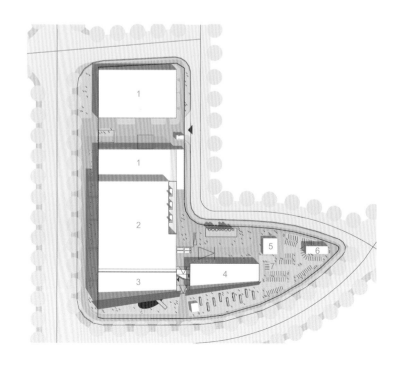

Architecture Envelope

The main façade of R&D building and the south façade of creation centre make the architecture spread along Xinhua Road. The solar photovoltaic panel wall rises from the earth in a concise and strong way. Photovoltaic panels of different colours compose the company's logo on the wall, creating a huge business card for the company, quite visually striking. Along the west, the red logo has a wonderful architectural effect, echoing with the red bridge at the entrance, achieving a perfect unity in exterior design. The main plant area uses stainless frame and green envelope to wrap the architecture. The bamboo green construction also makes the large-scale architecture present an inspiring charm of South China.

The Application of Renewable Energy and Energy Conservation Technology

The architecture uses Suntech Power's own product – Solar photovataic panel to replace conventional glass wall and to function as envelope the the main façade. The photovoltaic panels will function as a power station for office/R&D building and recreation centre. The energy for illumination and hot water will come from the inexhaustible solar power. The exterior walls of the plant are all made of stainless frame. The green envelope which consists of plants wraps the building. It not only embellishes the building, but also creates a micro eco-enviroment, reducing the energy assumption and making the interior comfortable in summer and winter. The design reflects the application of renewable energy source, showing a long-term development tendency of efficient green architecture. The application of Ground Source Heat Pump, on the base of application of solar energy, further advances the renewable application in the architecture, finally achieving the goal of "Zero-energy" building.

1. Full View
2. Façade
1. 全景图
2. 立面

一、概述

将建筑形态创造和节能技术充分结合，在创作中充分挖掘可再生能源的潜力，推进可再生能源在建筑中规模化的应用，实现建筑创能，成为本建筑设计重点思考的问题。而研发楼建筑超简洁的外观与室内空间的丰富灵动形式强烈对比成为建筑的最大亮点。

厂区按照生产区和办公辅助区分开设置，根据大宗物料运输需求和消防设置道路系统，整个建筑从功能上由生产楼、办公研发楼和康乐中心三部分组成，力求创造出一个造型新颖、环境优美的现代化企业的形象。

二、建筑形态

办公研发楼和康乐中心以复杂的功能及位置造型的重要性成为建筑首要部分。内部空间各自形成一个单层的高大空间。研发楼朝向共享空间分别按功能设置七个楼层，每个楼层功能又分成不同的区域，通过靠近生产楼的公共走道区连成一体。将每个楼层各区域互相错动咬合，行成丰富的空间感。康乐中心将整个建筑内部形成一个整体：球场、健身房、餐厅等各功能空间独立于大空间中央，空间形态在变化中寻求统一，各功能间通过同一个空间互相交流，融会贯通。生产楼四层高度，建筑面积5.4万平方米，采用钢筋混凝土结构，以其大体量构成建筑主体。生产楼设计本着经济性、功能性为主的原则。以简洁、纯净的形象展示产品的高科技特性。

Model of Elevation 外立面模型图

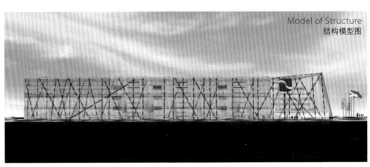

Model of Structure 结构模型图

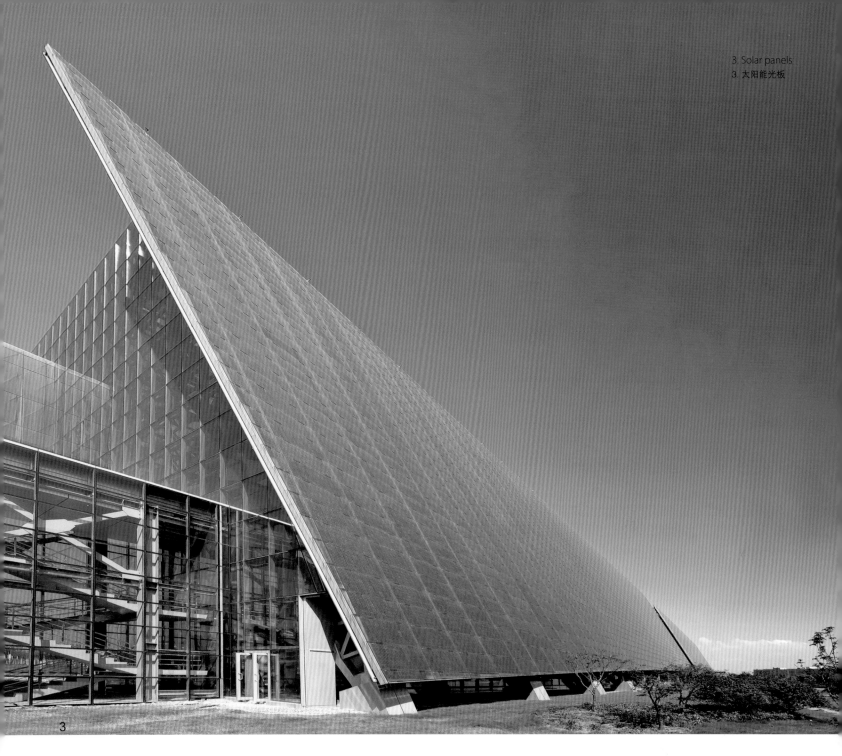

3. Solar panels
3. 太阳能光板

Cross Sections (Below): 横断截面图（下方两图）：
- Office 办公
- Recreation 康乐
- Transportation 交通
- Equipment 设备
- Air conditioner pipe 空调管井
- Production area 生产区
- Parking 停车
- Research room 研发

三、整个建筑外壳

主立面的办公研发楼和康乐中心南立面，整个建筑形象沿新华路充分展开，太阳能电池板墙以一种简洁有力的形象拔地而起，不同色彩的电池板组成的整墙公司Logo，成为企业巨大的商业名片，极富视觉冲击力。同时西侧沿路采用红色的公司标记丰富建筑效果，与入口处红色廊桥遥相呼应，在外形整体设计上达到完美统一。主体厂房部分外墙采用不锈钢架加绿色植物组成的绿色外衣将建筑包围，竹状的绿色构件使大体量建筑也体现出江南地区特有的清灵韵味。

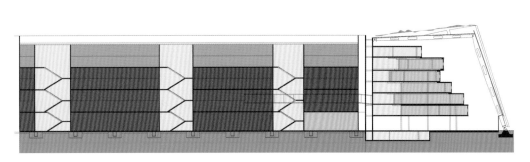

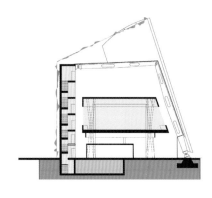

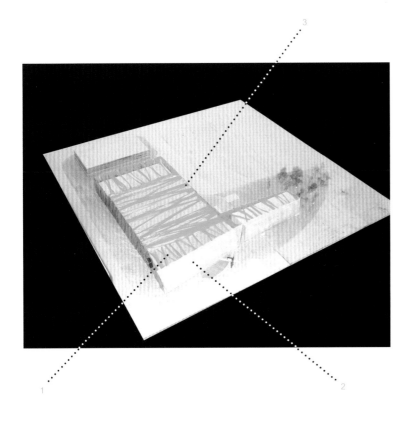

Skin (Left):
The head with its floating rooms cannot exist without any additional energy regulating measures. Therefore it will be surrounded by an "oxygen mask". This additional cover made of glass surrounds both the office building and the recreation centre and uses to full capacity the maximum volume of the building site.

1. *A greenhouse in the interspace*: An interspace like a greenhouse rises between the combined office/recreation and relaxation and can also be used for climatic balance.
2. *Power Station*: A solar power plant will be installed at the south side of the glass cover. Apart from producing energy, this huge panel which is composed of photo-voltaic cells acts as business card of the company.
3. *Second (Green) Skin*: The entire production hall as well as all the other parts of the vitreous cover will be wrapped with wire netting and ropes and will be planted with wisterias and wild vine.

The company's vision of "energising a green global future" is realised by this green energy building, because:
– a green skin will have arisen around the building within few months
– high technology for a global green future will be produced inside the building
– the office building produces its own electricity due to solar power plant

皮肤（左图）：
带有漂浮屋的建筑头部不能离开附加的能量调节控制装置而存在。因此，它必须有"氧气面具"。这个包围了办公楼和康乐中心的玻璃罩最大量的使用了建筑基地。

1. 中间的温室：中间部位联结办公楼和康乐中心的像一个玻璃覆盖的升起的温室。这区域可以休闲放松，也可以用来调节气候平衡。
2. 能源站：太阳能设备安装在南面的覆盖的玻璃上。除了生产能源，这面用太阳能光伏电池组成的巨大的板也可以用作公司的商业名片。
3. 第二层（绿色）皮肤：如同玻璃面覆盖其他部分一样，整个生产车间将采用金属丝网和绳包裹，并种上紫藤和野生葡萄藤。

在这个绿色建筑中，公司的"为地球绿色未来充电"的理念得到体现，这是因为：
——在几个月内，一层绿色皮肤将围绕建筑生长出来
——在建筑中，服务全球绿色未来的高科技将被制造出来
——办公建筑利用太阳能装置生产自己所需的电能

四、可再生能源利用及节能技术

建筑采用尚德自身产品——太阳能光电板取代传统玻璃幕墙，做为主立面外墙围护结构，做为办公研发楼和康乐中心的能源站，其照明、热水等能量都来源于取之不尽、用之不竭的太阳能。厂房外墙均采用不锈钢构架加绿色植物组成的绿色外衣将建筑包围，美化建筑同时使建筑围护结构形成生态小环境，冬暖夏凉降低能耗。充分体现了可再生能源利用和建筑形态设计的统一性，体现出长远的节能绿色建筑发展趋势。地源热泵系统的运用，在太阳能利用的基础上，进一步推进了可再生能源利用在本建筑中的实践，使研发办公楼最终实现"零能耗"目标。

Green Energy (Below):
The warmth and coolth production for component activation and climate ventilating systems and the hot-water preparation in the OFFICE BUILDING is carried out via Green Energy.

The energy is produced 100% from geothermal energy collectors under the foundation plate of the PRODUCTION HALL, the OFFICE BUILDING and the RECRATION BUILDING as well as from the heat recovery of all climate ventilating systems and the waste gas heat recovery.

The energy for the HEATING PUMP is produced 100% from the Photovoltaic-Panel energy system.

绿色能源（下图）：
办公区的空调冷热和热水能量是通过绿色太阳能提供。

太阳能是100%产生于生产厂房/办公楼和康乐中心建筑集热板的地热能力，与空调系统和废气的热能同样有效。

热泵能量是100%产生于太阳能收集系统板。

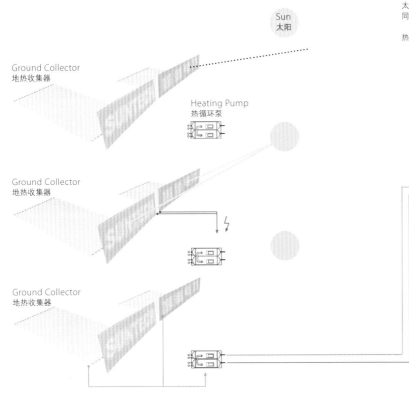

1. Outgoing air overflow
2. Air induction
3. Cooling water
4. Heating water

1. 向外的空气流
2. 进气口
3. 冷水
4. 热水

22,000m²h

4. Entrance
5. Lobby
6. Exhibition
7. Interior landscape
4. 入口
5. 大厅
6. 展示
7. 室内景观

Solar Power Station (Below):
1. Second green skin
2. Production hall
3. Office building
4. Vertical garden
5. Recreation
6. Conceptual model: office building with stratums

太阳能能源站（下图）：
1. 第二绿色皮肤
2. 生产区
3. 办公建筑
4. 垂直花园
5. 康乐中心
6. 概念模型：层叠状的办公建筑

Diagram of the Concept (Left):

The Concept
Body+Head+Skin=Building
Referring to the concept, the production hall forms the main body of the building. Both the office building and the recreation centre, kind of grow out of the main part like a head.

1. *Body:* The new production hall will be attached to the existing one in the south of the building. With its four floors the new hall forms a compact structure.
2. *Head:* The office building and recreation centre will be built out of single-storey organic stratums, which grow out of the new production hall in the south. The plan of each stratum, each floor, depends on different requirements concerning structure and surface of the offices, different needs of light as well as on the recreation programme.

Due to different floor plans on each storey, a vibrant dynamic arrangement, an organic landscape, with partly even floating rooms emerges by piling the stratums one upon the other. Terraces which can be reached from the above storeys complete this very special experience.

Identity and emotive quality:
The new building composes a landscape which shows not only its emotional but also its architectural uniqueness.

概念分析图（左图）：

概念：
躯干+头部+皮肤=建筑
参考这个概念，生产车间构成了建筑的主体躯干。办公楼和康乐中心则像生长在躯干外的头部。

1. 躯干：新的生产车间南面延伸出来的办公楼和康乐中心将被建造成有机的层叠状，每一层的设计，取决于结构的要求、办公及康乐中心对灯光的不同要求。
2. 头：由于每一层不同的平面，甚至部分飘浮的房间一层层的堆积，展现出一种充满生气的、动态的排列，一种有机的景观。

识别性和动人的特质：
新建筑形成的景观展现了它动人的、建筑风格上的独特性。

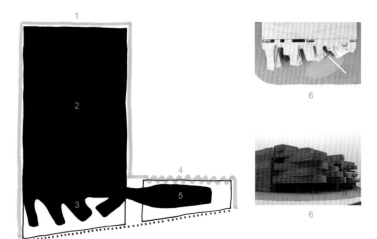

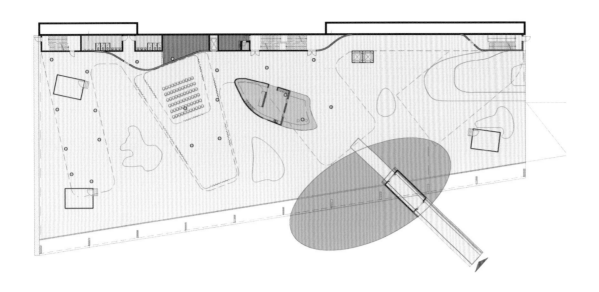

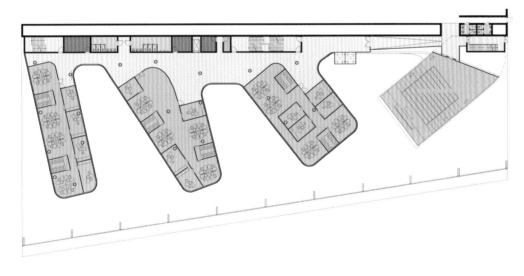

Office level 1 (Above), 2 (Left) and 6 (Bottom) Plan:
Office
Recreation
Transportation
Equipment
Restroom
Air conditioner pipe
Storage
Water
Green Skin

办公室一层（上图）、二层（左图）和六层（底图）平面图：
办公
康乐
交通
设备
卫生间
空调管井
储藏
水
绿色表皮

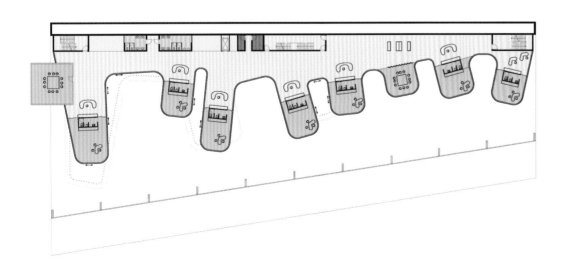

053

KUNSHAN NUCLEIC ACID SCIENCE AND TECHNOLOGY PARK
Kunshan, Jiangsu Province
logon

昆山小核酸产业科技园行政中心

江苏省 昆山市
罗昂建筑设计咨询有限公司

Site Area: 1.14ha
Gross Floor Area: 13,447m²
Completion Time: 2011
Architect: logon
Photographer: logon
Client: Tsinghua Science Park Kunshan
Cost: 3,200yuan/m² (including façades and equipment)
Energy Consumption: 150-170kWh/m²/year
Plot Ratio: 1.00
Greening Rate: 39.5%

占地面积：1.14公顷
建筑面积：13447平方米
建成时间：2011年
建筑设计：罗昂建筑设计咨询有限公司
摄影师：罗昂建筑设计咨询有限公司
业主：昆山清华科技园区
造价：3200元/平方米（包括外立面和设备）
能耗：150-170千瓦时/平方米/年
容积率：1.00
绿地率：39.5%

Kunshan Nucleic Acid Science and Technology Park – A Low-cost Sustainable Architectural Example

The new Kunshan Nucleic Acid Science and Technology Park is located at the intersection between Yuanfeng Road and Gucheng Road in New Hi-tech Development Zone, Kunshan, which is the first intersection on the boundary between Suzhou and Kunshan. The administrative building facing the main road is a landmark architecture. As the first to-be-built architecture, this administrative centre will become a flag to welcome guests.

The administrative centre is a platform for visitors and staff to communicate. The six-storey building is mainly the public service area, with over-ground construction area reaching 11,400 square metres, with reception, meeting rooms, leisure reading rooms, exhibition area and administrative office area spreading in the lower three floors. The flexible office area is located on the top floor and also opens to companies in the park and suppliers outside the park. The under-ground construction area is 2,050 square metres, mainly as parking and storage spaces.

Design Approach
First of all, the designers from logon identify three major design goals, and then confirm the design approaches. First, to build a landmark architecture at the gateway of Kunshan; Second, to provide a landmark public facility in Kunshan Nucleic Acid Science and Technology Park; Third, to establish a design criterion for the future development of the site.

Sustainability and energy saving are not the specific design requirements. The budget is 3,200 yuan/m² – a medium investment. Therefore, the budget for special design or green architectural design is limited. The problem is how to make the building a energy-saving architecture with an energy assumption of 150 kWh/m²/year.

In the earlier period of the design, logon adopts a unique design method: based on the existing conditions, it adopts sustainable design strategy and learns from both domestic and oversea experience to use internal tools and energy-saving standards. Besides, the design team holds the belief that only sustainable design is a successful design.

What is sustainability? The concept contains three "E"s – Economy, Equity and Ecology.
Economy means being practical and feasible, with high return benefits;
Equity means the project could fulfill all the user's requirements;
Ecology means it can save energy and reduce the carbon footprint.

According to sustainable concept, logon made design strategies as follows: First of all, the landmark architecture of Kunshan's gateway doesn't demonstrate itself through height or complicated exterior design, but through a clean design: a mysterious solid masonry structure facing the intersection, which is connected in the middle with a large glass atrium. This

Bird's-eye View Rendering
鸟瞰效果图

design is both visually impressive and energy-saving. By using less than 60% exterior glass and setting only a few entrances at east and west sides, the cost is lowered and the isolation is ensured.

Secondly, the administrative centre demonstrates its landmark status with some details: as a science and technology park, convenience in use, flexibility in space and a long-term accurate position are important indicators for high-quality design. As scientists, the architects focus on all details and materials both inside and outside the building. The interior climate and working environment is also important: users could enjoy natural landscape through different windows and fresh air would provide the staff with a great experience, both good for the work efficiency.

Last, the administrative centre sets a quality standard for Kunshan Nucleic Acid Science and Technology Park: the clean and elegant exterior form combines efficiency, function, high interior air quality and low energy assumption. The materials, volumes, entrances, energy-saving standards and landscape design will provide effective guide for the future architecture designs on this site.

LEC

The reason why logon cooperates with LEC is that compared with other certifications, LEC has two obvious advantages. First, LEC has sustainable design software co-developed by Chinese and German experts. It learns from the two countries' experience and is specifically designed for China's climate and environment. Second, the software can help to build energy-saving effects on the first phase of the design. With this smart design software, logon successfully reduced 50% of the energy assumption.

Up to now, Kunshan Nucleic Acid Science and Technology Park is the first project that obtains LEC. It obtains 2-star of winter and 4-star of summer. Without any increase in budget, the actual assumption is only 150-170 kWh/ m²/year.

Materials and Details

All the materials come from China. The architects made a strict selection in price, energy-saving effect and quality. The unique artificial stone finishing panel deserves to be mentioned. This 2.1×1.025m large-scale finishing panel is the only way to show the clean and unique façade. All the other top-level materials are used appropriately in the places they belong. Through LEC design software, logon optimises the design effect. Since the exterior sun-shading boards and expensive energy-saving glass can contribute little to the overall energy-saving effect, the architects chose standard double glass as a substitute and put the spare money into the isolation equipments on

Site Plan
总平面图

the wings. In this way, the quality of the interior space where people would spend most of their time is highly improved. The atrium connecting two wings has enhanced the overall ventilation condition. Simple sun-shading equipments are installed on the windows. The architects also pay special consideration to details to produce an effective and simple construction. At last, with the professional guide from BBS International, the air-conditioning and ventilation have achieved an optimised effect.

1. Building façade, panel and glass effect
2. Perspective Southwest view
3. Street view, south
4. Glass façade, building foyer
1. 建筑物立面、面板和玻璃的效果
2. 西南立面
3. 南立面街景
4. 玻璃幕墙、建筑门厅

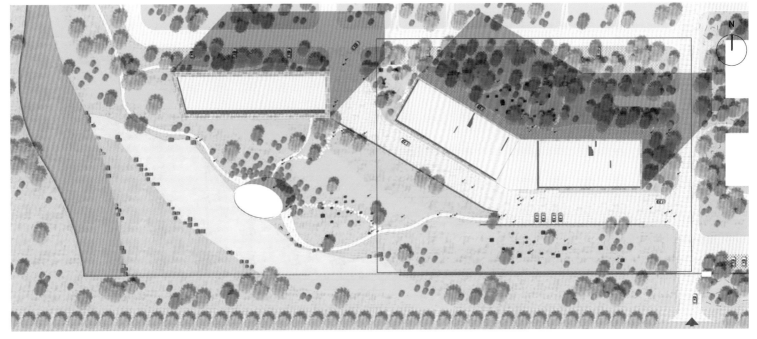

Landscape Design of North of Architecture (a and b)
- Make the eco botanic community at the north of architecture;
- Create the micro-topography based on earth balance, to enrich vertical space;
- Make a leisure space in the open area surrounded by the forest, and add some sculpture inside there;
- Use winding footways to connect all the function spaces;
- Use the grass brick at the car parking area, bicycle parking area and service channel to extend the green area;
- The sight of the car parking area, bicycle parking area and garage is blocked by the eco botanic community.

Landscape Design of South of Architecture (c)
- Use grass and a few trees which have high branches;
- The ground is covered with mixed grass to ensure the grass is evergreen in four seasons;
  The grass is resistant to trampling, and this will increase the scope of activities;
- The metasequoia trees are a background to contrast the restaurant building;
- Pool near the restaurant is covered with pebbles,
  and the shallow water can be used as activity space at the same time.
- Some sculpture-like seats on the grass can be used and appreciated.

建筑北侧景观设计（a和b）：
- 建筑北侧营造植物生态群落
- 确保土方平衡为前提营造微地形景观，丰富竖向空间
- 林荫环绕的开敞处点缀以雕塑式座凳，提供休闲场所
- 蜿蜒小径将各个功能空间联系起来
- 地面停车位、自行车停车场、服务通道处用植草砖铺砌，增大绿化面积
- 利用植物群落对停车位、自行车停车场、地下车库出入口进行视线遮挡

建筑南侧景观设计（c）：
- 建筑南侧以草皮为主，少量搭配分枝点高的乔木，保证视线开阔
- 草皮选用混播草种，确保草皮四季常绿，耐践踏，增加活动范围
- 用水杉林作为背景烘托餐厅建筑
- 靠近餐厅建筑的景观水面选用卵石作为池底
  浅水面，兼做活动空间
- 耐践踏草皮上散布雕塑式座椅，观用两顾

3

昆山小核酸产业科技园行政中心——一个低成本可持续建筑的范例

全新的昆山小核酸产业科技园行政中心位于昆山高新开发区元丰路古城路交界路口，这是苏州和昆山市边界的第一个交叉路口。行政中心大楼直面公路主干道，是这里的标志性建筑。同时，它也位于昆山小核酸产业科技园区的最重要的入口处。作为这里即将建成的第一栋建筑，行政中心大楼成为园区迎接客人的一面旗帜。

行政中心大楼是访客和园区员工沟通交流的平台。6层高的建筑主要是公共服务区域，地上建筑面积达到11,400平方米，前台接待区、会议室、休闲阅览室、展览区以及行政办公区集中分布在一至三层。灵活办公区域位于顶楼，同时对园区内公司和园区外供应商开放。地下建筑面积为2,050平方米，主要作为停车场和储藏空间。

设计方法

罗昂建筑设计咨询有限公司首先明确了三个主要目标，进而确定设计方法。目标一，在昆山市入境处打造地标建筑。目标二，为昆山小核酸产业科技园提供标志性公共设施中心。目标三，为该地块未来发展制定设计准则。

可持续性和节能并不是该项目的特别设计需求。而且造价预算是每平方米3200元人民币，属于中等水平投资，特殊设计或者绿色建筑设计的成本空间十分有限，如何将它打造成能耗仅为150千瓦时/平方米/年的节能建筑呢？

罗昂在项目初期采用了一种独特的设计方法：以项目状况为基础，采取可持续的设计策略，借鉴以往国内外相关项目的经验，运用内部工具和节能建筑标准。除此之外，设计团队始终秉承一个理念：只有可持续设计才是成功的设计。

Office Space: Clear Height   办公空间：净高
Open office   开放办公室
Meeting room   会议室
Individual office   独立办公室

Technical Installations: 技术设备：
Air condition   空调设备
Electricity   电
Telecommunication   电信设备

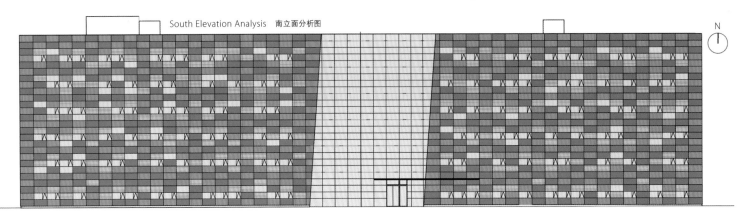

South Elevation Analysis 南立面分析图

Concept: 概念示意：
1. Economy   1. 经济
2. Equity   2. 公平
3. Ecology   3. 生态

Master Plan: Parking 总平面：停车
- Car parking -35c   地面停车（35辆）
- Parking vip - 5c   贵宾车辆（5辆）
- Temporary parking - 10c 临时停车（10辆）
- Underground - 35c   地下停车（35辆）
- Bicycle parking -260b   自行车（260辆）

Master Plan: Traffic Analysis 总平面：交通分析
- Car parking   车流
- Service circulation 服务车流
- Underground access 地下车库入口
- Drop-off   下客区
- Service entrance   服务入口

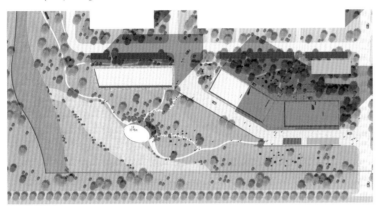
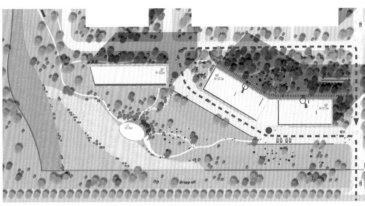
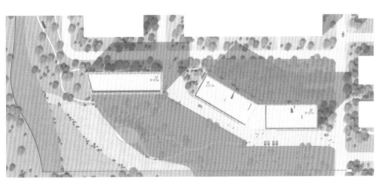

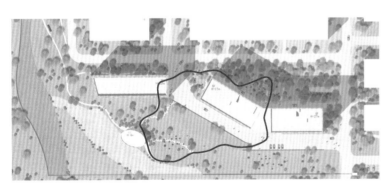

什么是可持续呢？可持续的概念涵盖了3个"E"（经济Economy、公平Equity、生态Ecology）。

经济——切实可行，回报收益高。
公平——项目满足所有用户需求。
生态——节约能源并减少碳排放。

根据可持续的概念，罗昂的设计策略如下：
首先，昆山市入口的地标性建筑不是由高度或花哨设计来表现的，而是通过简洁的设计来呈现：一个封闭神秘的大体量石材建筑面向十字路口，中间由一个大型的玻璃中庭连接。这个设计表现力极强，而且能够节约很多能源。采用少于60%的外立面玻璃，在东、西方向仅设几个出入口，在降低成本的同时，保证了隔热效果。

其次，行政中心还以建筑细节体现其标志性地位：作为一个科技园区，使用便利性、空间灵活性以及长期准确定位都是高质量的重要指标。建筑师如同科学家，认真地关注建筑室内外的每一处细节及材质。室内气候和工作环境亦十分关键：通过不同楼面窗户可以领略到自然美景，新鲜空气时时流通都为园区员工提供了良好的使用体验，有易于提高工作效率。

最后，行政中心大楼还为昆山小核酸科技园区设定了质量标准：简洁优雅的外部造型兼顾了高效性、实用性、高质量室内空间和低耗能。外立面材质、体量、入口状态、节能标准和景观规划一系列设计准则，为该地块未来其他建筑单体的设计提供了有效指导。

可持续设计认证（LEC）

罗昂和可持续设计认证机构（LEC）合作的理由是，相较于其他认证，它有两个显著优势：第一，可持续设计认证机构（LEC）有德中两国专家联合开发的可持续设计软件，吸取了两国专家的经验，并完全针对国内气候环境。其次，可持续设计软件能够帮助建筑师在设计初期优化建筑的节能效果。罗昂通过这个智能设计软件，成功减少50%的建筑能耗。

MATERIALS FAÇADE (Above)
Variation of Four Materials: The façade has two different stone panels and two different glass units. The combination of the different elements creates a balance between light and dark, sun and shadow.
Sunshaded Glass:
a. Stone - grey
b. Dark grey
c. Stone - double glazing
d. Silk-screened glass
e. Laminated glass with high reflection
f. Thermo isolated glass

立面材料（上方7图）
四种不同的材料：运用两种石材与两种玻璃窗，这两种材质的组合创造出明与暗、光与影的平衡。
遮阳玻璃：
a. 灰色石材
b. 深灰色石材
c. 双层玻璃
d. 印刷玻璃
e. 反光玻璃
f. 隔热玻璃

昆山小核酸科技产业园行政中心是目前国内最早获得可持续设计认证（LEC）的项目：获得了冬季2星认证和夏季4星认证。在不增加任何预算的情况下，现实能耗仅为150–170千瓦时/平方米/年。

材料和细节

这个项目所有材料都产于中国，建筑师针对价格、节能效果和质量进行严格挑选。其中人造石外装饰板十分独特，值得一提。这个尺寸为2.1米×1.025米的大型装饰板是体现简洁独特立面造型的唯一手段，其他所有材料也都是顶级质量，都被恰到好处地运用到需要的地方；罗昂通过LEC设计软件达到设计效果最优化；室外的遮阳板和价格不菲的节能玻璃实际上对整个建筑的节能效果影响很小，所以最后选择标准双层玻璃作为替代，并把省下的资金投入到增加大楼左右两翼的隔热设备。这样一来，人们停留时间最多的室内空间质量大大提高；连接左右两翼建筑的玻璃中庭改善了整个自然通风状况；简单的遮阳设备安装在嵌入的窗户上；建筑师还特别注意细节使施工过程变得简单高效；最后，在莱默建筑设计工程咨询有限公司的专业指导下，空调和通风等设备的使用均达到最佳效果。

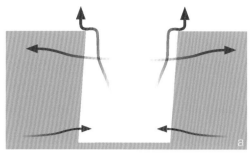

SUSTAINABILITY: CLIMATE PRINCIPLES (a, b, c and d)
Natural Ventilation: Use the geometry of the building to create a natural movement and exchange air. The central atrium creates a chimney effect transporting the air through the whole building and out through the atrium.
Sunshading: Control of the solar impact on the room temperature by alternating the use of regular glass and sun protection glass, silk – screened glass or high reflective glass.
Insulation: High quality, thick and efficient wall insulation to keep cooling and heating costs low. Additionally the façade of the atrium is a double glazed façade, which prevents it to become overheated.

可持续发展：小气候设计准则 (a, b, c 和 d)
自然通风：利用建筑的几何形状使空气自然流动，交换。中庭创造了一种烟囱效应，空气通过整个办公空间最后由中庭排出。
遮阳：利用正常玻璃和外面玻璃例如印刷玻璃、反光玻璃、双层玻璃结构来控制室内温度。
隔热层：厚的、高质量和有效的隔热层将有利于减小能耗。同时中庭立面也是双层玻璃结构避免建筑过热。

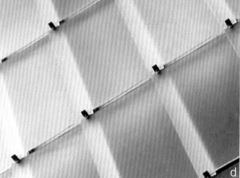

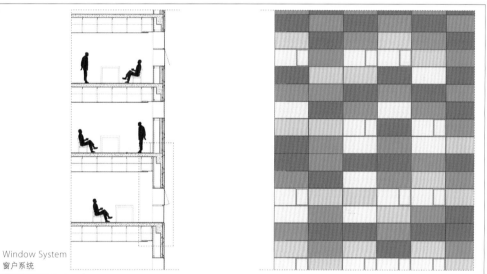

Window System
窗户系统

MATERIALS: INTERIOR (Below)
Comfort and Durability: Outstanding materials providing a comfortable, dynamic and modern environment.
a. Metal mesh
b. Dark grey steel frames
c. Wood panels
d. Black stained concrete

室内材料（下方4图）
舒适耐久：高品质的材料创出出舒适、现代又充满着活力的环境。
a. 金属网
b. 深灰色钢框架
c. 木板
d. 黑色杂点混凝土

Rendering
效果图

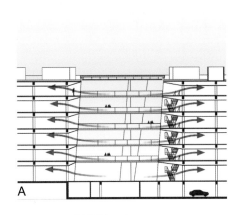 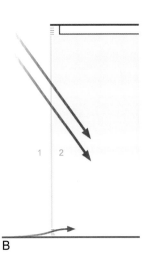 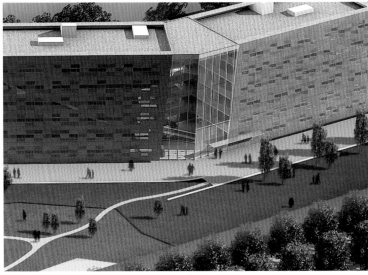

CLIMATE: WINTER (Above)
Natural sun heating: The façade of the atrium heats up by the sun from south in cold but sunny winter days. The warm air will naturally flow into the open floors and help to warm up the inside climate.
1. Cold Exterior
2. Warm Interior
A. From the atrium warm air heated by the sun flows into the office spaces
B. Sun warms up the glass façade

小气候设计准则：冬季（上方两图）
*自然太阳光热：*在寒冷但太阳充足的冬天，中庭幕墙随着太阳光照南面温度逐渐升高。热空气将通往各层，升高室内温度。
1. 外面冷
2. 里面热
A. 被阳光加热的空气流向各层办公空间
B. 阳光使立面变热

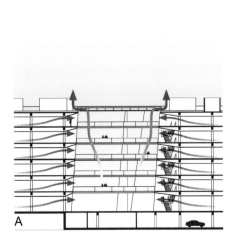   

CLIMATE: SUMMER (Above)
Natural ventilation: The central atrium cuts through all six floors, which gives a chimney effect, meaning that warm used air will be transported through the whole building and out through the atrium. Additionally the façade of the atrium is a double glazed façade, which prevents it from become overheated.
1. Warm Interior
2. Cold Exterior
A. From the office spaces on each floor warm used air flows out and up through the atrium
B. Zoom: Glazed façade with internal sun shading
C. Atrium without sun shading
D. Atrium with sun shading

小气候设计准则：冬季（上方两图）
*自然通风：*中庭六层通高，提供了类似于烟囱的效应。也就是说，暖风将通过整幢建筑并由中庭排出。并且，中庭玻璃幕墙是双层结构，可以避免建筑过热。
1. 里面热
2. 外面冷
A. 热空气经过各办公空间汇聚到中庭，向上排出
B. 放大：双层玻璃与室内遮阳
C. 不采用遮阳设施的中庭
D. 采用遮阳设施的中庭

Basement Floor Plan:
1. Entrance
2. Track
3. Parking
4. Fire basin
5. Pumping
6. Elec
7. Tlecom
8. Refrigerator
9. Ventilation
10. Diffuser
11. Safe entrance
12. Detoxification
13. Duty room
14. Ventilation (war)
15. Toilet (war)
16. Drink (war)
17. Wall (closed for war)

地下一层平面图:
1. 出口
2. 车道
3. 车位
4. 消防水池
5. 泵房
6. 变配电
7. 通信
8. 冷冻机房
9. 风机房
10. 扩散室
11. 密闭通道
12. 滤毒室
13. 防化值班室
14. 战时进风机房
15. 干厕
16. 战时设XXT玻璃钢饮用水箱
17. 临空墙临战封堵

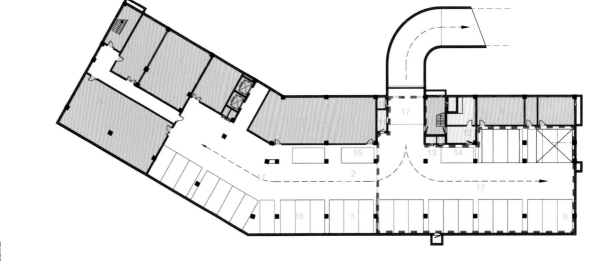

Ground Floor Plan:
1. Lobby
2. Reception, shop
3. Café library
4. Dining area
5. Café/shop
6. Other supportive area and circulation
7. Exhibition area
8. VIP/lounge area
9. Conference

一层平面图:
1. 入口门厅
2. 接待、商店
3. 咖啡书屋
4. 餐厅/多功能厅
5. 咖啡/小卖部
6. 其他服务及交通空间
7. 展览
8. 贵宾室
9. 150人会议中心

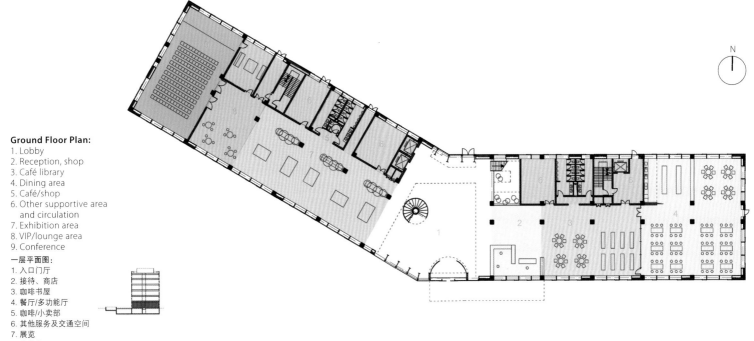

063

5. Vertical view in the foyer
6. Internal glass façade in the foyer
5. 建筑门厅,水平垂直视图
6. 建筑门厅,内部的玻璃幕墙

Rendering
效果图

STRUCTURE (Below Two)
Column-beam structure: In order to provide a clear open space with a high variation of possible division, a column-beam structure is used with larges spans. The columns are placed asymmetrically in the middle and in the façade, to give as deep as possible span in the south and less deep rooms towards north. In the middle the absence of beams gives place for installations.

结构设计准则（下方两图）
梁柱结构体系：为了提供灵活开敞的空间和分割的灵活性，设计中我们运用大跨度的梁柱体系。梁柱不对称地分布在楼层中央和幕墙区域。从而使南面有尽可能大的进深，同时控制朝北的进深，中央区域没有过梁，方便布置管线。

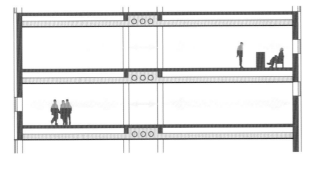
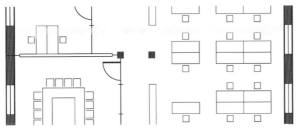

Rendering
效果图

QINGPU ENVIRONMENTAL MONITORING STATION
Shanghai
Atelier Liu Yuyang Architects

上海青浦环境监测站

上海市 青浦区
刘宇扬建筑事务所

Site Area: 5,000m²
Design/Completion Time: 2009-2010/2011
Architect: Atelier Liu Yuyang Architects
Local Design Institute: Shanghai City Architectural Design Co., Ltd.
Façade Consultant: Kighton Façade Consultants Co., Ltd.
Lighting Consultant: Unolai Design
Photographer: Jeremy San

占地面积：5000平方米
设计/建成时间：2009–2010年/2011年
建筑设计：刘宇扬建筑事务所
合作设计院：上海都市建筑设计有限公司
幕墙顾问：凯腾幕墙设计公司
灯光顾问：十聿照明设计有限公司
摄影师：Jeremy San

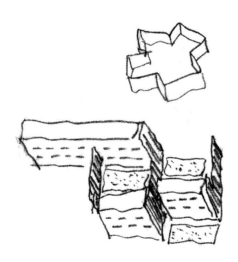

This design combines the Western "scientific view of nature" and the ancient Chinese "humanistic view of nature" and through an individual's own observation of nature and intuitive experience of surrounding environment, forms a "phenomenal view of nature". The design not only satisfies criteria of scientific functional layout and humanistic nature of space and form, but utilises the visual moment when one is in architecture to constitute the inherent relationship between man and natural environment, and it adopted the "three walls, three courtyards and three floors" approach to blend the space in between wall and courtyard, and formed the entity of the architecture. This proposal focuses not on the style of architecture, but on the relationship between architecture and landscape (gardens and nature), and the visual moment of transitional space between buildings.

Energy and Ecological System

The project adopts a geothermal heat pump system for normal heating and cooling, and through energy exchange via a series of 80-metre-deep underground pipes, the energy consumption of the building is reduced. The steel cable sunscreen system allows local vines and creepers to grow along the building façade and reduces the indoor temperature in summer; in addition, the cables are suitable for birds to rest, the birds' droppings naturally form organic fertilisers for the vegetations below. The entrance canopy incorporates reverse beams to allow for planter spaces on top, effectively reducing heat gain below and providing unique landscape views above. The outdoor gardens with natural grass and permeable paving design, with intermittent paving and soil allow for maximum rain water retaintion underground.

本案设计融合了西方的"科学自然观"和古老中国的"人文自然观",并透过个人自身对自然的观察、直观式的体验周遭环境,形成一种感染人们内心的"现象自然观"。在满足功能布局的科学性和空间形式的人文性之后,本案利用建筑所捕捉到的视觉瞬间,构成人与自然环境的内在关系。本案建筑采取了"三墙、三院、三楼"的手法,糅合了墙与院的空间,并形成了建筑的形体。苏轼《望江南·暮春》中的"春未老,风细柳斜斜"一语道出了江南文化的根源——它那温徐的气候和细致的环境。本案最终的着墨之处不在于建筑本身的形式趣味,而在于建筑与景观(园林、自然)的关系,和建筑之间的过渡空间的视觉捕捉。

能源生态系统说明

项目的日常供暖制冷采用了地源热泵系统,透过80米深的垂直管井与地下恒温进行能量交换,形成了有效建筑节能策略。建筑立面通过外挂金属丝索的蔓藤遮阳系统来降低夏日室内温度。此外,外挂钢索适合鸟类栖息,而鸟类的排泄物又为蔓藤提供自然施肥,形成了有机循环。在建筑入口的屋檐通过反梁结构形成二层室外的植栽空间,降低屋檐底下的室外温度,也为二层提供特殊景观效果。同时,室外庭院采用了自然植被以及青砖与土壤的间隙排列铺装,形成了自然的可渗透铺地并能最大程度地将雨水保留于土壤之下。

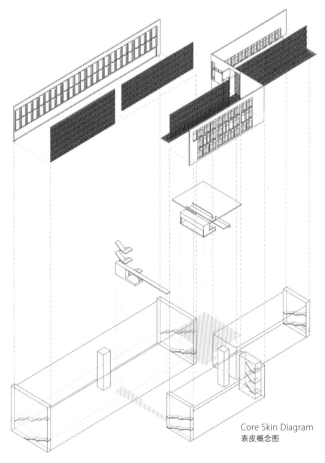

Core Skin Diagram
表皮概念图

1. The north elevation of laboratory building at night
2. The south elevation of office building
1. 实验楼北立面夜景
2. 办公楼南立面

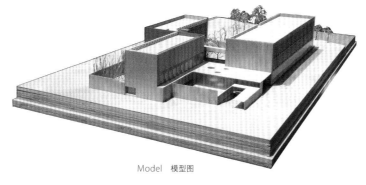

Model 模型图

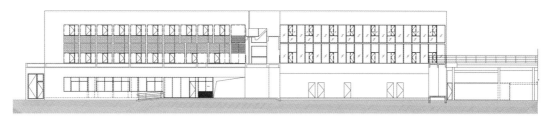

Elevations 立面图

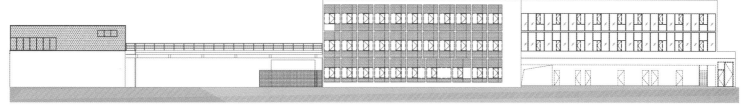

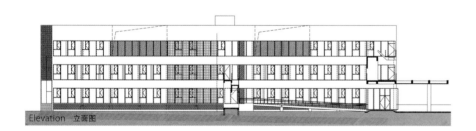
Elevation 立面图

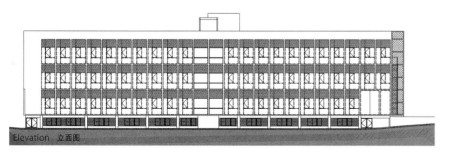
Elevation 立面图

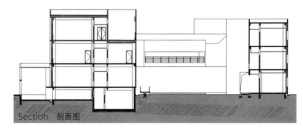
Section 剖面图

3. The north elevation of laboratory building
4. Entrance of laboratory building
3. 实验楼北立面
4. 实验楼入口处

Level -1 (Right):
1. Garage
2. Pump house
3. Substation
4. Storage
5. Refrigeration storage
6. Preparation room
7. GSHP room

地下一层平面图（右图）：
1. 车库
2. 水泵房
3. 配电间
4. 仓库（戊类）
5. 冷库
6. 准备间
7. 地源热泵房

5. View of entrance courtyard (stone court)
6. View of laboratory building entrance and central courtyard (tree court)
5. 入口庭院（石院）景观
6. 实验楼入口及中央庭院（树院）景观

Level 1:
1. Air supply room
2. Gas fire suppression
3. Large equipment room
4. Atomic absorption room
5. GC-MS
6. Standard gas room
7. Changing room
8. Preparation room
9. LC room
10. IC room
11. Reagent room
12. Mercury vapourmeter
13. Sample pretreatment
14. Large equipment preparation
15. Men's room
16. Women's room
17. Computer room
18. Washing and disinfection room
19. Instrument room
20. Office

一层平面图：
1. 气源室
2. 气体灭火
3. 大型仪器室
4. 原子吸收室
5. GC-MS
6. 标气室
7. 更衣室
8. 准备间
9. 液相色谱室
10. 离子色谱室
11. 试剂室
12. 测汞仪
13. 样品预处理
14. 大型仪器准备间
15. 男卫
16. 女卫
17. 网络机房
18. 消解间
19. 仪器室
20. 办公室

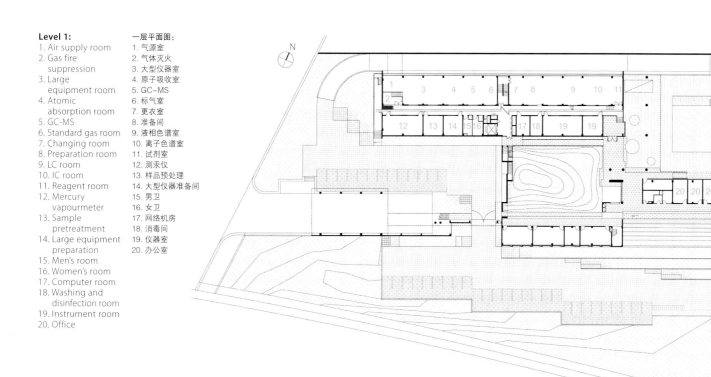

Level 2:
1. COD
2. LAS
3. Chemistry laboratory
4. Atmosphere analysis room
5. Oil laboratory
6. BOD
7. Reagent room
8. Landscape earthing terrace
9. Heat source room
10. Preparation room
11. Scale room
12. Colorimetric room
13. Chemistry decomposition room
14. Digestion room
15. Laboratory water making room
16. Plenum room
17. Osphresiometer
18. Restaurant
19. Office
20. Accounting office
21. Archives
22. Bioanalysis
23. Chemical analysis
24. Instrumental analysis

二层平面图：
1. COD
2. 酚砷LAS
3. 化学试验室
4. 大气分析室
5. 石油实验室
6. BOD
7. 试剂室
8. 景观覆土平台
9. 热源室
10. 准备间
11. 天平室
12. 比色室
13. 化分室
14. 消解间
15. 实验室制水间
16. 配气室
17. 嗅辨室
18. 餐厅
19. 办公室
20. 财务室
21. 档案室
22. 生物分析
23. 化学分析
24. 仪器分析

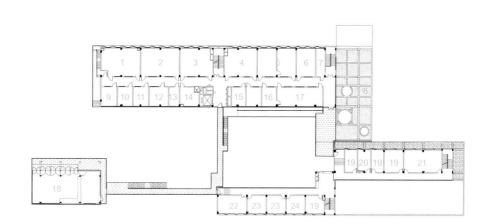

Level 3:
1. Biology laboratory
2. Automatic station
3. Preserved space
4. Large equipment room
5. Microorganism preparation room
6. Terrace
7. Men's room
8. Women's room
9. Storage
10. Reception room
11. Office
12. Meeting room

三层平面图：
1. 生物试验室
2. 自动站
3. 预留
4. 大型仪器室
5. 微生物准备间
6. 露台
7. 男卫
8. 女卫
9. 储藏
10. 接待室
11. 办公室
12. 会议室

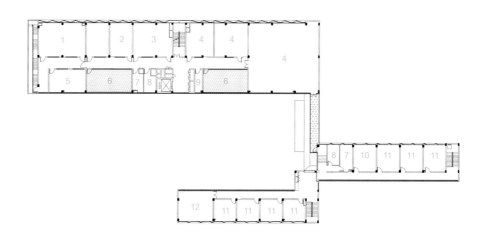

8

7. View of side courtyard (bamboo court)
8. Vertical green cable system
7. 边院（竹院）景观
8. 垂直绿化钢索系统

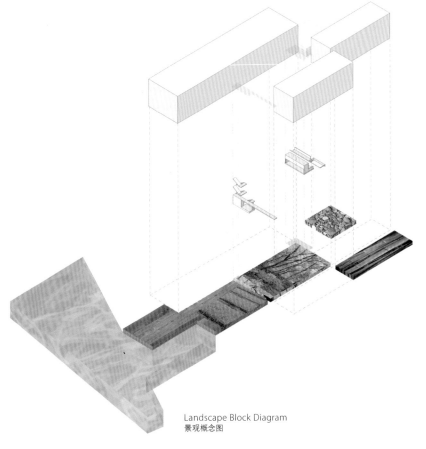

Landscape Block Diagram
景观概念图

Green Architecture Composite System Diagram:
1. Birds and mesh
2. Green façade and apron
3. Outdoor sun desks
4. Geothermal mechanical room
5. Geothermal heat pump tube well
6. Rainwater permeable pavement
7. HVAC (heating ventilation air condition) pipes
8. Pebble garden
9. Canopy planting

绿色建筑综合系统图解：
1. 小鸟与网架
2. 绿色立面与散水
3. 室外露台
4. 地源热泵机房
5. 地源热泵预埋管井
6. 雨水渗透铺地
7. 通风空调管线
8. 卵石花园
9. 顶盖植栽

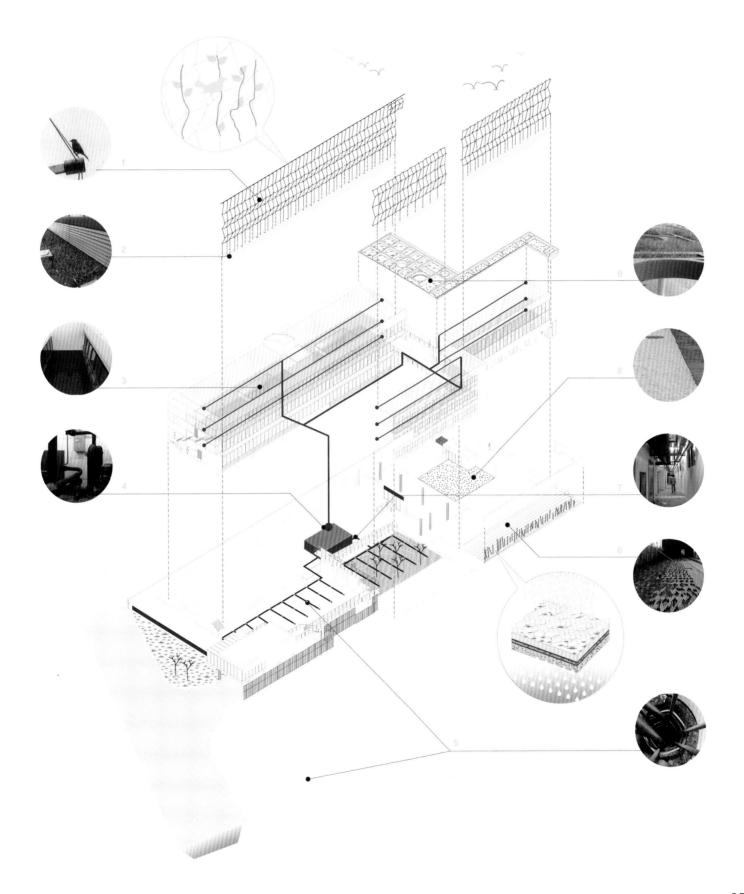

XI'AN JIAOTONG-LIVERPOOL UNIVERSITY
Suzhou, Jiangsu Province
Robert Goodwin, Michael Bardin / Perkins+Will

西安交通利物浦大学

江苏省 苏州市
罗伯特·古德温，迈克尔·巴丁 / 帕金斯威尔建筑师事务所

Gross Floor Area: 45,000m²
Completion Time: 2011
Architect: Robert Goodwin, Michael Bardin / Perkins+Will
Photographer: Eduard Hueber
建筑面积：45000平方米
建成时间：2011年
建筑设计：罗博·古德温，迈克尔·巴丁 / 帕金斯威尔建筑师事务所
摄影师：爱德华·胡博

Inspired by the famous water gardens in nearby Suzhou, this 45,000 sqm Integrated Science Campus for Xi'an Jiaotong Liverpool University proposes a new academic paradigm in which building and landscape interact to form an innovative environment for scientific learning and discovery. Instead of the traditional "quadrangle", the design adopts a strategy of horizontal and vertical layering to create a series of interwoven spaces that optimise movement, interaction and environmental response.

The primary design strategy locates high-traffic lecture halls at grade, where the ground plane rises to enclose them in a sculptural landscape of water pools, green walls, garden roofs, and continuous walking paths. Formed around and over this landscape array are a series of sinuous metal armatures that define courtyards and organise larger spaces on the site. Finally, classroom, laboratory and office spaces, clad in glass and ceramic tile, are inserted into the armatures to complete a distinctive expression of the integration of technology and nature.

Within this approach, a number of important strategies are emphasised. Classrooms and offices are modular to permit future adaptability; laboratories are designed as flexibly partitioned "lab lofts" that can accommodate changes in teaching or research as institutional needs evolve. Faculty offices provide direct access to teaching and research spaces; multiple entries, exterior stairs, lounges and circulation links create an interactive framework for a variety of campus experiences. Ultimately the design supports a pedagogical strategy of "active learning" that encourages interactive, collaborative activities both inside and outside the classroom.

The design responds directly to its environment: the classroom wings are oriented to optimise solar orientation, with limited exposure on the eastern and western sides and substantial sun-shading to protect south faces.

Natural daylighting is emphasised through narrow floor plates and large amounts of glass to reduce the energy consumption required for artificial lighting. In addition, all classrooms and lounge spaces have operable windows to allow for passive ventilation and to further reduce energy needed for heating and cooling. Rainwater is collected for both irrigation and reuse for non-contact uses in the building and for water features within the courtyards. Green roofs are used extensively in the lower levels on both horizontal and inclined surfaces to absorb stormwater and reduce heat gain.

Site Plan
总平面图

受到苏州著名水景园林的启发，西安交通利物浦大学综合科学校区创造了全新的学术建筑典范——建筑与景观相互融合，形成了创造性的科研环境。该项目设计摒弃了传统的"四合院"式的建筑结构，采用了水平与垂直层叠的形式，编制出运动、互动和环境响应的最优空间。

该项目主要的设计策略反映在人流量最多的阶梯教室。底层地面上升，以水池、绿墙、花园屋顶以及连续走道的雕塑景观将阶梯教室围住。这个环绕景观阵列的是一系列蜿蜒的金属骨架，它们划分出了庭院，并且在场地上组织出了更大的空间。最终，被包围在玻璃瓷砖结构中的教室、实验室和办公空间嵌入了金属骨架之中，形成了目前我们所看到的技术与自然相结合的独特景象。

设计还突出了许多重要的策略。教室和办公室采用了模块化设计，便于未来的改造；实验室被设计成灵活的"实验LOFT"，可以根据教学研究的需求进行改造。教职工办公室直接通往教研空间；多重入口、外部楼梯、休息室区和路线连接共同打造了交互式校园体验框架。设计支持了"主动学习"教学法，鼓励教室内外的互动与合作。

设计直接对环境作出响应：教学楼的朝向优化了日光朝向，东西两侧暴露很少，南侧设有大量的遮阳设施。狭窄的楼板和大量玻璃装配保证了自然采光，减少了人工照明的能源消耗。所有教室和休息空间也都设有可控式窗户，保证了被动通风，进一步减少了供暖和制冷的能源需求。雨水被收集起来用于灌溉，或再利用于楼内非接触用水与庭院水景设施。底层建筑水平和倾斜表面大量采用绿色屋顶，能吸收雨水径流并减少热增量。

1. Façade detail
2. Distant view of façade
3. Close view of façade
1. 外立面细节
2. 远景
3. 近景

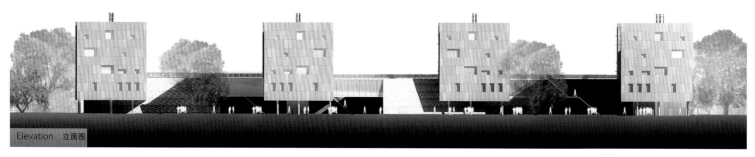

Elevation 立面图

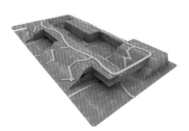

A continuous interior circulation loop at the first floor links the upper levels to one another and the ground plane.
二楼连续的室内循环回路将上层空间与底层连接起来

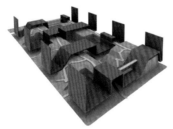

The supporting armature of the classroom and laboratory wings of the building are formed and supported by the lecture hall and landscape system below.
教室和实验室辅助设施由下方的阶梯教室和景观系统组成并支持

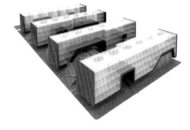

Building programme spaces and building systems fit within the envelope created by the lab and classroom wing armature.
建筑功能空间和建筑系统与实验室和教室所形成的表皮十分契合

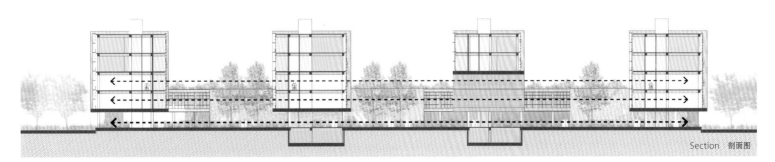

Section 剖面图

4. Roof garden
5. Night view
4. 绿色屋顶
5. 夜景

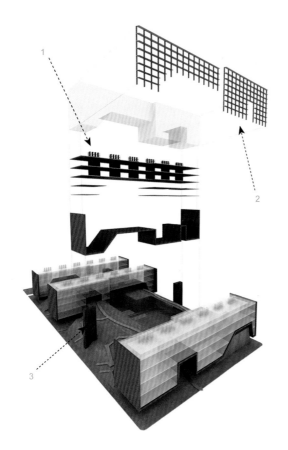

Building Systems and Solar Shading (Left):
1. Educational space fits within the building armature of each wing. The top two flexible laboratory floors are served by a vertical mechanical system which vents at the roof level.
2. Daylighting and solar gain are optimised by a system of fins and sunshade at each exterior envelope of the classroom wings, administration areas, and other occupied spaces.
3. Stair and lift cores rise from the ground floor to provide barrier-free access to all levels and spaces in the building.

建筑系统和太阳能挡板（左图）：
1. 每个翼楼内部都设有教学空间；顶楼的两层灵活实验室空间采用垂直机械系统，屋顶配有通气孔。
2. 自然采光和太阳能供暖通过翅片系统和外部遮阳系统得到了最佳调节，包括教室楼、行政区以及其他区域。
3. 楼梯和电梯核从一楼上升，为所有楼层和空间提供无障碍进出。

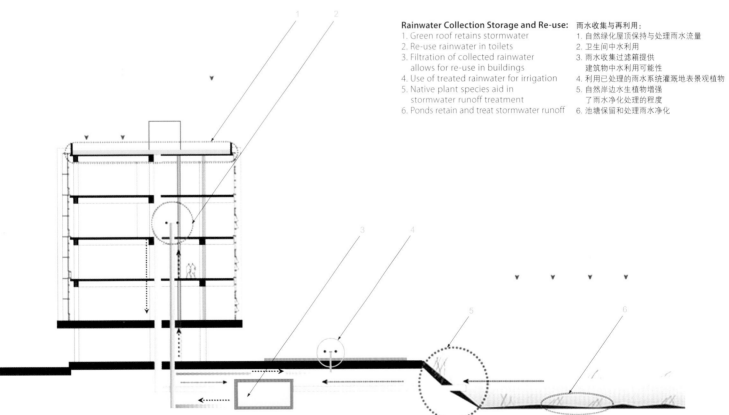

Rainwater Collection Storage and Re-use:
1. Green roof retains stormwater
2. Re-use rainwater in toilets
3. Filtration of collected rainwater allows for re-use in buildings
4. Use of treated rainwater for irrigation
5. Native plant species aid in stormwater runoff treatment
6. Ponds retain and treat stormwater runoff

雨水收集与再利用：
1. 自然绿化屋顶保持与处理雨水流量
2. 卫生间中水利用
3. 雨水收集过滤箱提供建筑物中水利用可能性
4. 利用已处理的雨水系统灌溉地表景观植物
5. 自然岸边水生植物增强了雨水净化处理的程度
6. 池塘保留和处理雨水净化

6. Study area
7. Lecture hall
6. 自习室
7. 阶梯教室

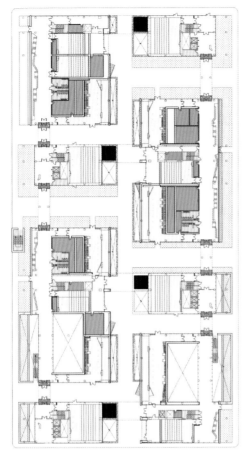

Ground Floor Plan
一层平面图

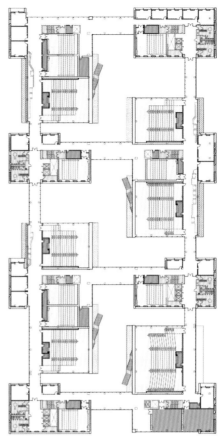

Second Floor Plan
二层平面图

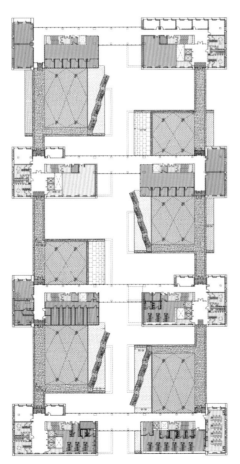

Third Floor Plan
三层平面图

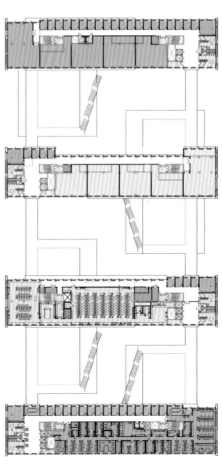

Fourth Floor Plan
四层平面图

RENOVATION OF NO.9 XINGHAI STREET, SUZHOU INDUSTRIAL PARK
Suzhou, Jiangsu Province
Suzhou Institute of Architectural Design Co., Ltd.

苏州工业园星海街9号改造工程

江苏省 苏州市
苏州设计研究院股份有限公司

Site Area: 18,000m²
Gross Floor Area: 12,673m²
Completion Time: 2010
Architect: Suzhou Institute of Architectural Design Co., Ltd.
Design Team: ZHA Jinrong, CAI Shuang, WU Shuxin
Photographer: ZHA Zhengfeng
Award: 2011 Award of Excellence of Jiangsu Urban-Rural Construction System

占地面积：18000平方米
建筑面积：12673平方米
建成时间：2010年
建筑设计：苏州设计研究院股份有限公司
设计团队：查金荣，蔡爽，吴树馨
摄影师：查正风
奖项：江苏省城乡建设系统建筑工程优秀设计一等奖

The total site area is approximately 18,000 square metres. Before renovation, the plant is a single-level concrete frame structure, appearing as a square of 80X80 metres. Therefore, the central day-lighting and ventilation are undesirable. After renovation, the whole architecture can be divided into three layers of space: inner courtyard, office and veranda rest space.

Combining the natural condition and existing environment of Suzhou, the architecture reserves 95% structure of the original plant, changing the existing 6,800-square-metre, 8.4-metre-high single-level industrial plant into a 12,673-square-metre green eco creative space. The designers also provide a gym and badminton court for the staff to relax and exercise in the 14-metre-high plant space.

The large-scale openable floor-to-ceiling glass around the building improves the natural ventilation. The designers used computer simulation analysis software to study the natural ventilation model and set up a reasonable design strategy. After utilising reasonable ventilation strategy, the interior temperature can be reduced by 2 to 3℃ compared with non-natural ventilation. On average, it reduces interior extreme temperature time by 570 hours, increases interior comfortable time by 720 hours and reduces air-conditioning time by 160 hours.

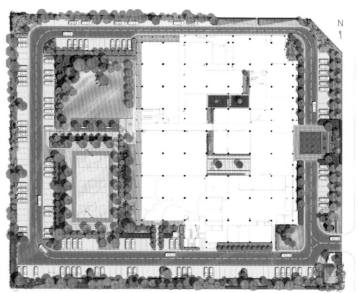

Site Plan 总平面图

The existing exterior solid wall has been transformed too. The glass windows provide plenty day lighting. The 11 skylights are transformed into operable power sunroofs, both improve interior ventilation and ensure the lighting requirement. Based on the large area and the lower height, the office building uses daylight illumination system. The system uses high-efficient light pipe and UV-filter diffuser to introduce daylight around the building into the interior. The first-floor corridor, hallway, open office area and conference room all depend on daylight in the day. Nearly no additional artificial lighting is needed, thus reducing the electricity assumption to minimum.

The upper part of the veranda is made of aluminium alloy shading grids. The east and west façades are added with vertical wood sun shadings. Climbing plants such as Campsis grandiflora and Chinese wisteria can climb through the columns and sun shadings into the horizontal metal grids, creating an eco shading. In winter, when the leaves fell off, the interior will have enough sun light. The exterior landscape around the balcony has several movable strip planters.

The exterior wall uses heat preservation materials. The main colour tone is white. White paintings have 90% high reflectivity, which will reflect most part of the heat radiation. The exterior windows use heat-insulation aluminium and hollow insulation glass to reduce the air-conditioning load. The green climbing plants on the west wall will also improve the insulation effect.

1. Ecological shading
2. The east elevation
3. Close view of the northeast corner
1. 生态外遮阳
2. 东立面景致
3. 东北角近景

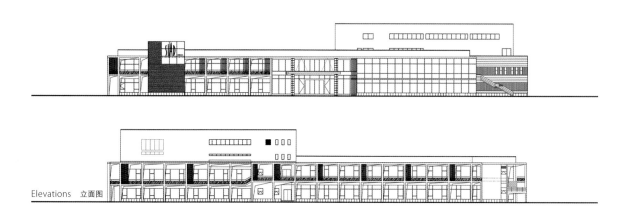

Elevations 立面图

Courtyard Section A:
1. Electromechanical design department
2. R&D reserved
3. Corridor
4. Internal courtyard
5. Conference room
6. Conference room
7. Training room
8. Exhibition room

Courtyard Section B:
1. Storage
2. Fitness centre
3. Interior design department
4. R&D reserved
5. Corridor
6. Corridor
7. CAD centre
8. Restaurant
9. Civil engineering department
10. Internal courtyard
11. Office
12. Corridor
13. Graphic centre
14. Office
15. Corridor
16. Dean's office
17. Balcony

Courtyard Section C:
1. Architectural design department
2. R&D reserved
3. Internal courtyard
4. Training room
5. Corridor
6. Conference room
7. Internal courtyard
8. Restaurant
9. Drawing and document room
10. Conference room
11. Corridor
12. Civil engineering department
13. Corridor
14. HR department

庭院剖面图A：
1. 机电设计部
2. 研发预留
3. 走廊
4. 内院
5. 会议室
6. 会议室
7. 培训室
8. 展览室

庭院剖面图B：
1. 库房
2. 健身中心
3. 室内设计部
4. 研发预留
5. 走廊
6. 走廊
7. CAD中心
8. 餐厅
9. 土建设计部
10. 内院
11. 办公室
12. 走廊
13. 图文中心
14. 办公室
15. 走廊
16. 院长办公室
17. 阳台

庭院剖面图C：
1. 建筑设计部
2. 研发预留
3. 内院
4. 培训室
5. 走廊
6. 会议室
7. 内院
8. 餐厅
9. 图档室
10. 会议室
11. 走廊
12. 土建设计部
13. 走廊
14. 人力资源部

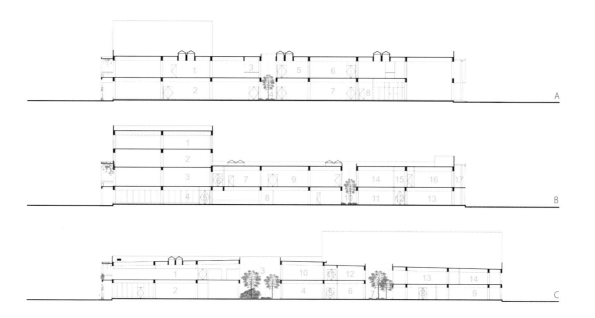

087

4. The west elevation
5. Snowscape
6. The south elevation
4. 西立面
5. 雪景
6. 南立面

厂区总用地面积约1.8万平方米。改造前厂房为单层混凝土框架结构，建筑近似80米×80米的正方形，中部采光与通风效果不理想。改造后整个建筑由内至外可分为三重空间——内庭院、办公空间和外廊休息空间。

建筑结合了苏州的自然条件和既有现状，完全保留了旧厂房95%主体结构，并通过在外部加设绿化外廊、加层、内部开挖内庭院等处理，使原面积6,800平方米、单层高8.4米的工业厂房，成为总面积为12,673平方米的绿色生态创意空间。在原厂房局部14米高的空间里，还设计了供员工休闲锻炼的健身房和羽毛球场。

建筑外围设置了大面积的可开启落地玻璃改善建筑的自然通风条件。运用计算机模拟分析软件，对建筑的自然通风模型进行研究，制定了最为合理的设计方案。在采用合理的自然通风措施后，室内温度可比非自然通风条件下降低2~3℃，平均减少室内极端温度时间570个小时，室内满意小时数增加720个小时，减少空调开启时间平均约为160小时。

原建筑外围的实墙也进行了改造，综合运用外墙上的玻璃窗采光。原屋顶的11个天窗被改造：设置了可开启电动天窗，改善内部的空气流通，同时保证空间的照明需要。由于办公楼占地面积大、层数少，办公楼照明采用了日光照明系统，该系统可通过高效光导管和带有紫外线滤除功能的透光罩将屋面四周的日光引入室内，二层的走道、门厅、大开间办公室、会议室等白天利用自然采光，最大限度的节约了照明用电消耗。

建筑外廊上部为铝合金遮阳隔栅。建筑东西立面还增加了竖向木饰遮阳板。种植的凌霄、紫藤等攀缘植物可通过柱子和竖向遮阳，攀爬至顶部的水平金属隔栅上，形成覆盖型生态遮阳。冬季植物叶子脱落后给室内带来充足的阳光。阳台外围一圈结合景观，还设置了可移动式条形种植槽。

建筑墙体外维护采用自保温的墙体材料，主色调为白色。白色涂料有90%的高反射率，可反射大部分热辐射。外窗采用断热铝合金型材和中空隔热玻璃，减少了空调负荷。西侧外墙增加的绿色攀缘植物也强化了隔热保温效果。

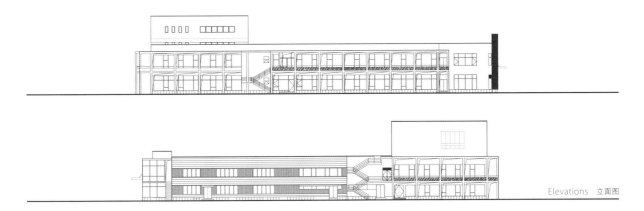

Elevations 立面图

7. Verandah detail
8. Balcony green shading
7. 外廊局部
8. 阳台绿化遮阳

Multi-Layer Shading (Right):
1. Dark grey metal coping
2. 30mm extrusion molding insulation panel
3. Fixed with steel nail and sealed by factice
4. Additional waterproof layer on the corner
5. Steel structural grids
6. Planting soil
7. Dry-hang granite finishing

复层生态遮阳（右图）：
1. 深灰色金属压顶
2. 30厚挤塑保温板
3. 钢钉钉牢油膏密封
4. 转角处做附加防水层
5. 钢结构隔栅
6. 种植土
7. 干挂花岗岩饰面

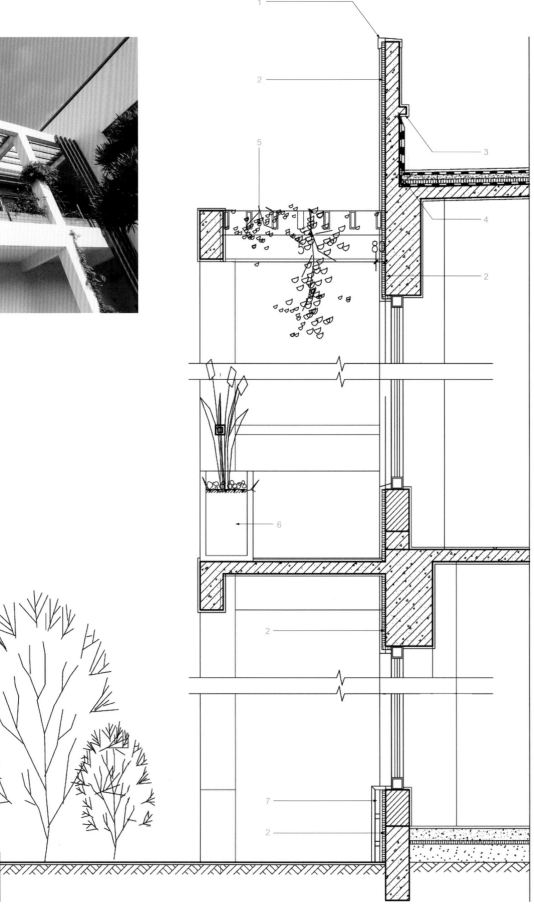

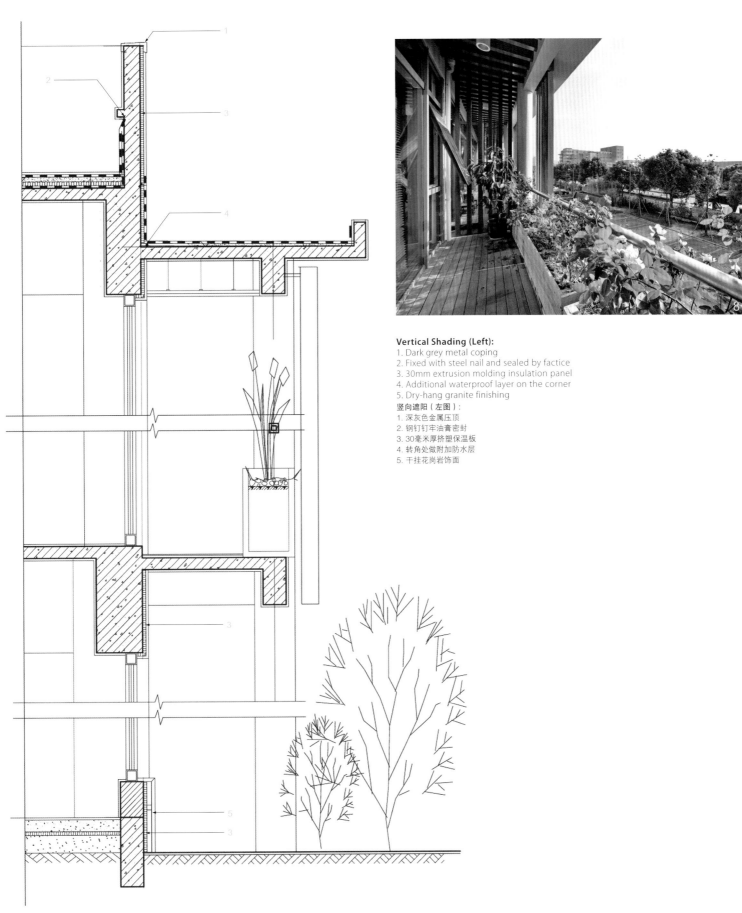

Vertical Shading (Left):
1. Dark grey metal coping
2. Fixed with steel nail and sealed by factice
3. 30mm extrusion molding insulation panel
4. Additional waterproof layer on the corner
5. Dry-hang granite finishing

竖向遮阳（左图）：
1. 深灰色金属压顶
2. 钢钉钉牢油膏密封
3. 30毫米厚挤塑保温板
4. 转角处做附加防水层
5. 干挂花岗岩饰面

Double Glazed Wall (Below):
1. Dark grey metal coping
2. 30mm extrusion molding insulation panel
3. Dry-hang granite finishing

双层玻璃幕墙（下图）：
1. 深灰色金属压顶
2. 30毫米厚挤塑保温板
3. 干挂花岗岩饰面

9. Ecological shading
10. Fixed shading
11. Ecological shading

9. 生态外遮阳
10. 固定外遮阳
11. 生态外遮阳

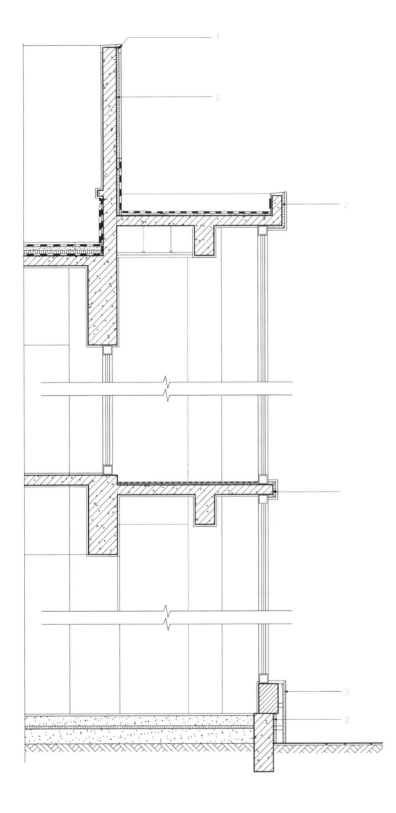

12. Garden
13. Inner courtyard
14. The rain-recovery landscape pool
12. 庭院
13. 内庭院小景
14. 雨水回收景观水池

A Variety of Green Eco Technology (Left):
1. Solar water heating system
2. Green roof
3. Cooking fume purification system
4. Internal courtyard
5. Double glazed system
6. Electronic openable skylight
7. Veranda eco shading system
8. Daylight illumination system

多种绿色生态技术（右图）：
1. 太阳能热水系统
2. 屋顶绿化实践区
3. 厨房油烟净化处理系统
4. 内庭院
5. 双层玻璃幕墙系统
6. 电动开启天窗
7. 外廊生态遮阳系统
8. 日光照明系统

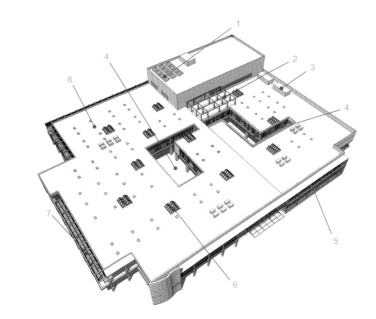

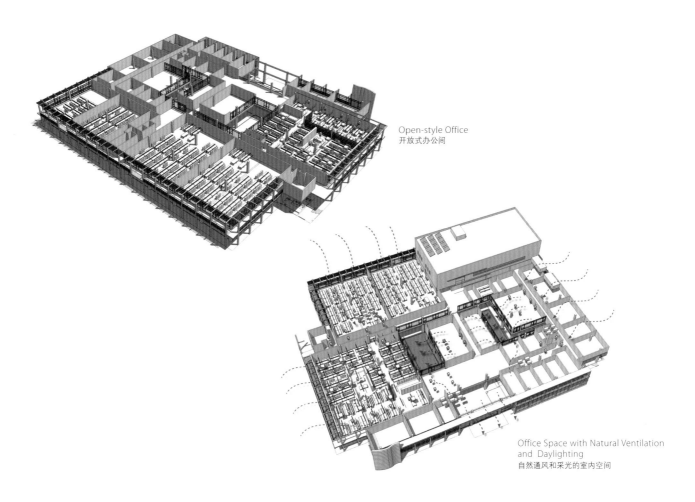

Open-style Office
开放式办公间

Office Space with Natural Ventilation and Daylighting
自然通风和采光的室内空间

15. The landscape pool making use of rain recovery
16. Hall space
17. The conference room making use of daylighting by skylight
15. 利用回收雨水的景观水池
16. 门厅
17. 利用改造天窗采光的会议室

18, 19. Roof ecological green
18、19. 屋顶生态绿化

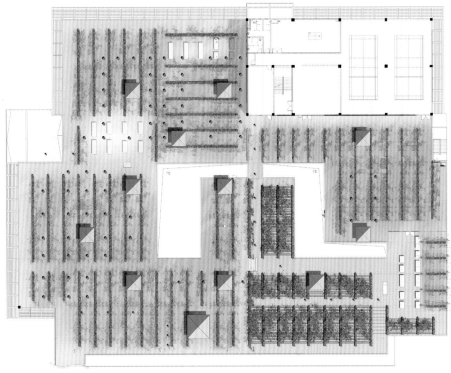

Roof Greening　屋顶绿化

Ground Floor Plan:
1. R&D reserved
2. Power distribution area
3. Restaurant
4. Drawing and document room
5. Internal courtyard
6. Landscape pool
7. Property management mode
8. Office
9. Conference room
10. Training room
11. Slide showroom
12. Graphic Centre
13. Marketing Department
14. Lobby
15. Exhibition room
16. Green building
17. Reconstruction company

一层平面图：
1. 研发预留
2. 配电区
3. 餐厅
4. 图档室
5. 内院
6. 景观水池
7. 物业管理模式
8. 办公室
9. 会议室
10. 培训室
11. 晒图间
12. 图文中心
13. 市场部
14. 门厅
15. 材料展示室
16. 绿色建筑
17. 重建公司

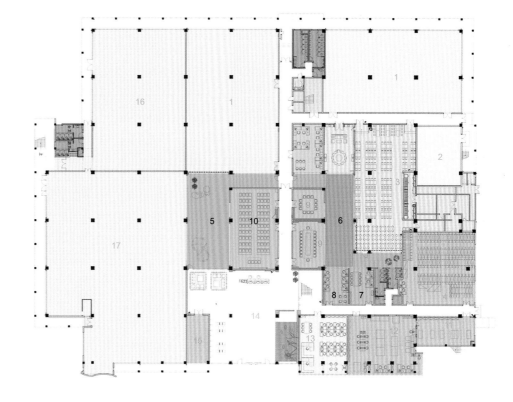

First Floor Plan:
1. Interior design office
2. Civil engineering department
3. Landscape design office
4. HR centre
5. Reference room
6. Financial department
7. General designer office
8. Office
9. Institute office
10. CAD centre
11. Printing room
12. Conference room
13. Electromechanical room
14. Structure department
15. Internal courtyard
16. Architecture department
17. Creation office

二层平面图：
1. 室内设计办公室
2. 土建部
3. 景观所
4. 人力资源中心
5. 资料室
6. 财务部
7. 总师办
8. 办公室
9. 院办
10. CAD中心
11. 打印室
12. 会议室
13. 机电部
14. 结构部
15. 内院
16. 建筑部
17. 创作所

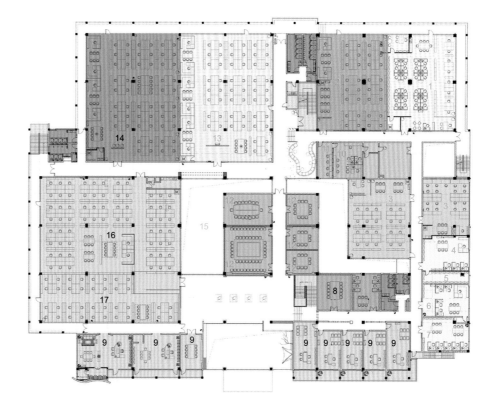

Openable Skylight (Right):
1. Openable skylight
2. Hidden electrical machines and wires
3. Coated tempered double glass window
4. Coated tempered double glass openable skylight
5. White painting
6. 30x30 wood-coloured aluminium alloy louvres

可开启天窗（右图）：
1. 可开启天窗
2. 电机电线暗铺
3. 镀膜钢化夹胶玻璃固定窗
4. 镀膜钢化夹胶玻璃可开启天窗
5. 白色涂料
6. 30×30木色铝合金遮阳百叶

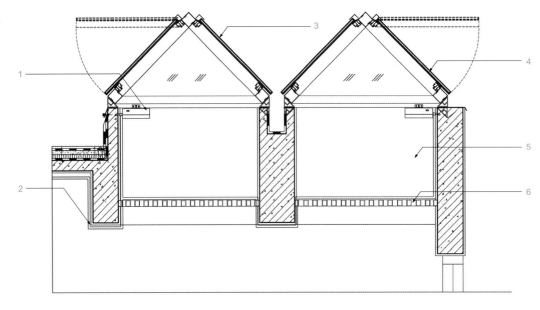

20. Sunlight lighting system
21. Office space making use of solar lighting system
20. 日光照明系统
21. 利用日光照明系统的办公空间

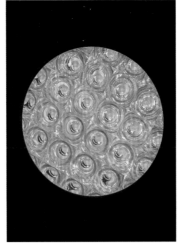

Nature - Sunlight
自然·阳光

Life - Vitality
生命·活力

THE GRAND PERGOLA – ADMINISTRATION CENTRE OF BIOLOGICAL OFFICE PARK SUZHOU
Suzhou, Jiangsu Province
WSP Architects

苏州生物纳米科技园
江苏省 苏州市
维思平建筑设计事务所

Gross Floor Area: 45,205m²
Design/Completion Time: 2006/2009
Architect: WSP
Principal: WU Gang, CHEN Ling, ZHANG Ying
Design Team: LIU Tao, HU Dawei, OUYANG Wei
Photographer: YAO Li

建筑面积：45205平方米
设计/建成时间：2006年/2009年
建筑设计：维思平建筑设计事务所
主设计师：吴钢，陈凌，张瑛
设计团队：刘韬，胡大伟，欧阳伟
摄影师：姚力

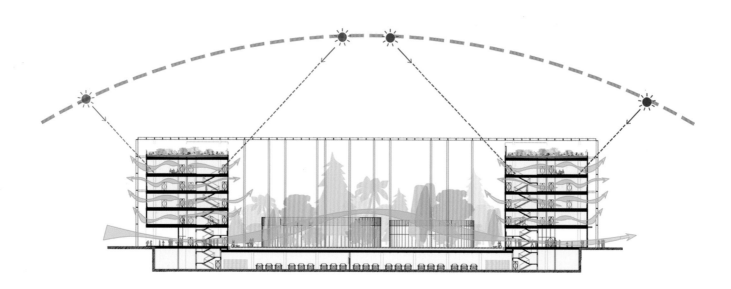

The Biological Office Park is located in Suzhou Industrial Park, which is a park for sustainable and ecological science and technology. The administration centre of the Biological Office Park is located in the western edge of the central park. The building consists of two parallel office slabs. The basic design intention of the administration centre is to create an open, ecological, sustainable and energy-saving office.

The space between two parallel office slabs creates an inner courtyard which is linked to the central park of the Biological Office Park. The administration centre is the visual focus point of the whole Biological Office Park. Circular design elements are applied in landscape design, the building façade and interior design. They resemble the idea of cells creating an impression of a breathing lung.

A metal structure with perforated metal panels as a grand pergola covers office slabs and the courtyard. The roof is a low-tech, inexpensive and effective method to effectively filter sunlight and fierce wind, thereby keeping the interior space comfortable. The grand inner courtyard under the pergola has enough height and airflow to allow plants to grow freely. Furthermore, vertical gardens are designed as integral part of office buildings, thereby achieving an ecological working environment. The ground floor is lifted up, connecting the outer park with the inner park.

The same perforated panels of the pergola are used on the exterior of the façade. Additionally the façade, windows and exterior doors are well insulated to minimise energy consumption. The whole structure is based on a grid of 1,050mm.

1. Grand pergola
2. Over view of façade
1. 巨型花架
2. 全景

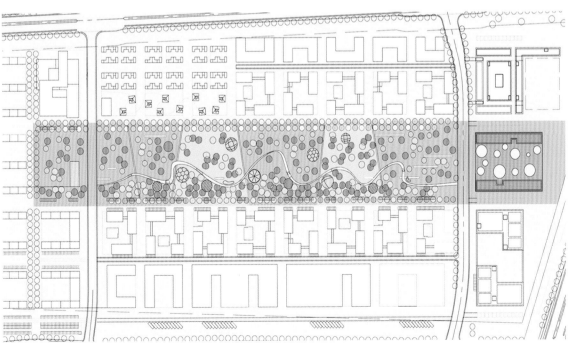

Site Plan
总平面图

3

苏州生物纳米科技园位于苏州市工业园区内，设计目标是创造可持续发展的生态智能化科技园区。当地传统的建筑和园林的关系被继承和革新性地再现，对于中国古典建筑如何与现代功能要求、现代建筑语言相结合的问题具有启发性的意义。对于使用者、参观者和周边邻居来说，它是现代建筑，同时又具有为他们所熟悉和认知的中国传统空间的意味。

生物园的管理中心位于苏州生物纳米科技园中央园的最西端，由两栋平行布置的办公建筑组成，本项目与中国现有的大量办公建筑截然不同，是一个具有生态化与可持续发展倾向的代表性的办公建筑，它的办公环境不再局限于简单封闭的办公空间，而是与周边环境融为一体。管理中心最基本的设计意图是创建开放式的、生态化的、具有可持续发展能力的节能办公空间。两座平行的办公建筑之间自然形成了一个园林广场，从而将中央公园引入管理中心，使得管理中心成为中央公园的有机组成部分，并在视觉上成为该区域的焦点。在建筑上以及建筑之间的园林上覆盖着技术简单、成本低廉、效能良好的遮阳篷，解决了园林广场和建筑立面的遮阳问题。建筑的一层架空，人们可以从建筑前广场自由进入这一有着遮阳篷的区域。源于"细胞"概念的圆形设计元素从中央公园一直延伸到管理中心的外立面和中庭建筑。整个管理中心犹如一个可呼吸的绿肺。

管理中心的内广场上空被建筑的多孔板景观盒所围合，支撑这个景观盒的是广场内若干两两交叉的柱结构，这些柱子在景观上寓意为交错的竹丛，为此，在广场中，设计种植了软的竹丛与这些硬柱结构相互呼应并弱化它们，这也使中心广场的景观与建筑更加融合在一起。

在建筑上以及建筑间的园林上覆盖着技术简单、成本低廉、效能良好的铝镁合金穿孔遮阳篷，是"花架"这一庭院景观元素的巨型化演绎，可有效过滤多余阳光和强风，使内部空气保持舒适状态，并为爬藤植物提供了生长的框架。遮阳篷下巨大的景观内庭具有充足的层高和良好的气流循环，使植物可自由生长，结合各楼层的空中花园形成绿色的办公环境。建筑的底部架空，人们可以从建筑前广场自由进入这一有着遮阳篷的区域。

除了使用穿孔板外立面进行第一层的保温隔热外，建筑还采用了挤塑聚苯外保温墙面，断桥铝合金门窗和中空玻璃等方式增强节能效果。整体建造基于1050毫米的模数。

3. Façade detail
4. Courtyard
3. 外墙
4. 庭院

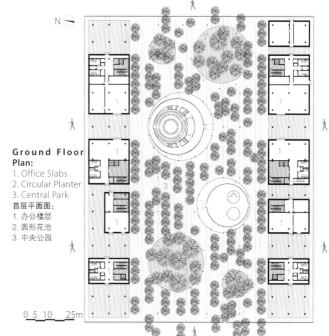

Ground Floor Plan:
1. Office Slabs
2. Circular Planter
3. Central Park

首层平面图：
1. 办公楼层
2. 圆形花池
3. 中央公园

5. Grand pergola
6. Green detail
7. Entrance of exhibition space
5. 巨型花架
6. 局部
7. 展厅入口

KPMG-CCTF COMMUNITY CENTRE
Cifeng Town, Pengchuan, Sichuan Province
The Oval Partnership Limited

毕马威安康社区中心

四川省 彭川市 磁峰镇
欧华尔顾问有限公司

Gross Floor Area: 450m²
Completion Time: 2010
Architect: The Oval Partnership Limited
Photographer: KMPG China
Client: CCTF, KPMG China

建筑面积：450平方米
建成时间：2010年
建筑设计：欧华尔顾问有限公司
摄影师：毕马威中国
业主：中国儿童少年基金会，毕马威中国

Building Rural Sustainability Through Green Innovation and International Collaboration

KPMG China, in partnership with China Children and Teenagers' Fund (CCTF) and Chengdu Women's Federation, have built an exemplar green community centre in Cifeng Village, Sichuan Province, China as part of efforts to renovate the earthquake-devastated poor region and promote the development of the local rural communities.

The architect has made full use of locally sourced materials in building a people-oriented, energy-saving, and environmentally sustainable village. The project's main construction materials have come from renewable resources, such as a full reconstituted bamboo post & beam structure prefabricated from local factory and sourced from local sustainable forest, pollution free agricultural straw fibre panel for wall and roof system, bamboo cladding and floor finish.

This low-budget philanthropy project has received pro bono manpower, resources and financial support from 30 international and domestic green enterprises, research institutions and government agencies through partnership.

The 450-square-metre Community Centre was opened on 17 May 2010, and is used for local children's extracurricular activities and villagers' vocational training. The project has advanced sustainable rural community development by means of corporate sponsorship, public participation and public-private partnership, and will serve as a paragon for improving sustainable construction, educational, cultural and recreational facilities in rural communities.

1, 2. Construction progress
3. Detail
4. Full view of façade
1、2. 施工过程
3. 特写
4. 全景

Elevations 立面图

Site Plan 总平面图

Building Highlights a Series of Rural Green and Intelligent Strategies

– programmes assisting companies, individuals, and charitable foundations reach out to imaginative social, environmental and community schemes in less developed areas in China with funds, services and volunteer effort
– passive bio-climate design to improve IEQ and achieve a low carbon scheme
– new moon-shape tall and gracefully curving single-storey structure that captures the varying hourly and seasonal angles of the sun and air flow effectively that additional heating, cooling and lighting are nearly unnecessary. Specifically, passive design measures include:
– best southeast solar orientation
– in summer sunlight shaded out of the building by south high eaves; in winter sunlight penetrating into heart of building through high windows
– roof solatube inducing natural lighting and balancing indoor lighting level
– landscape pergola shading on south
– reducing glazing area of north opening for less heat loss in winter
– curving roof assisting prevailing north wind
– cross and high window stack effect natural ventilation
– raised floor promoting natural ventilation and reducing site impact
– highly insulated cavity walls
– double glazing window with thermal-insulated laminated timber frame
– concrete floor providing thermal mass

Innovative Green Labelled Materials to Reduce Carbon Emission, Waste and Indoor Pollution

– reconstituted bamboo structure prefabricated from local factory and sourced from local sustainable forest
– research has shown that YBkouki and Dasso bamboo products used in the project are CO_2 neutral due to the fact that bamboo is an important and very fast CO_2 "fixator"
– formaldehyde free agricultural straw fibre panel for wall and roof system
– reconstituted bamboo cladding and indoor and outdoor floor finish
– recycled reconstituted timber for window

Other Green and Intelligent Features

– advanced communication connectivity and intelligent facilities for rural education
– integrate a 2000-year-old "the God of Earth Temple" within the design
– bamboo mortise-tenon joint structure for robust seismic performance
– energy-efficient measures such as LED lighting
– local rural training such as eco farm and green lifestyle
– native species planting and sustainable drainage for rural irrigation

5. North elevation
5. 北立面景致

Double-Arch Form on Plan and Section Embracing Nature and People:
1. Solar tube inducing natural lighting and balancing indoor lighting level
2. LED energy-efficient lighting
3. North high windows allowing natural ventilation and day lighting
4. Reducing the glazing area of north window to achieve less heat loss in winter
5. Raised floor to promote natural ventilation and reduce site impact
6. Concrete floor providing thermal mass
7. Curving roof assisting air flow
8. Cross ventilation and stack effect
9. South high windows inducing natural daylight and solar heat in winter, providing views as well
10. Landscape shading on south
11. Double glazing window with thermal-insulated laminated timber frame
12. Food production and sustainable drainage on community central plaza

连拱形的平面与剖面设计烘托出人与自然和谐相处的气氛：
1. 太阳能管引入了自然采光，平衡了室内照明水平
2. LED节能灯
3. 朝北的高窗保证了自然通风和采光
4. 减少北窗的玻璃区域以减少冬季的热流失
5. 高架地板提升了自然通风并减少了场地影响
6. 混凝土地面提供了蓄热体
7. 弧线屋顶促进了空气流通
8. 交叉通风和烟囱效应
9. 朝南高窗在冬季引入了自然采光和太阳热能，提供了良好的视野
10. 南侧景观遮阳
11. 双层玻璃窗，配有保温层叠木框
12. 社区中央广场的食品生产和可持续排水系统

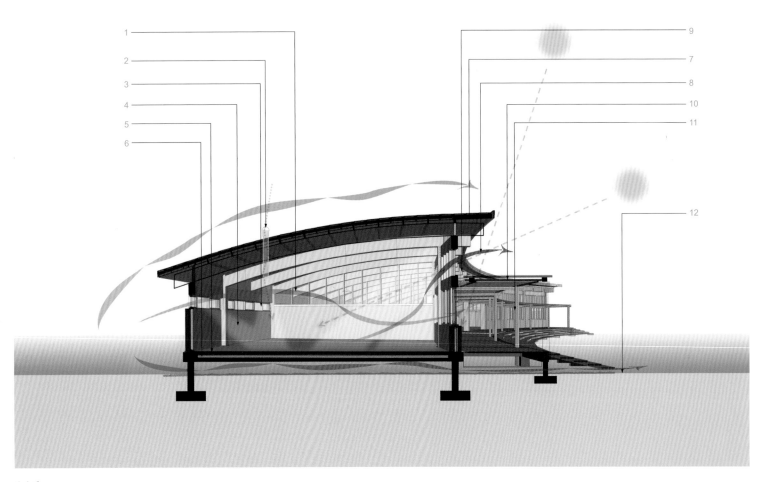

6. Entertainment room
7. Tea house
6. 社区娱乐中心
7. 茶室

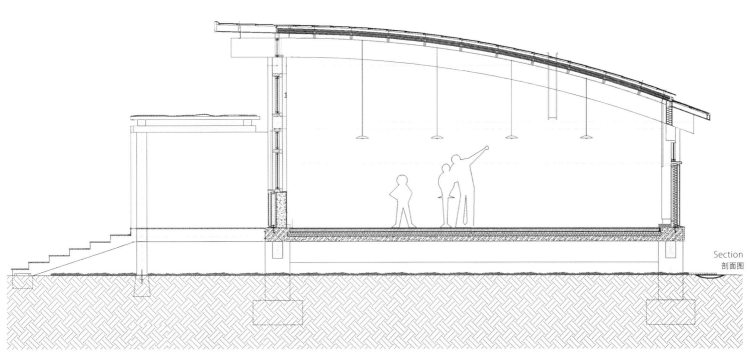

Section
剖面图

绿色创新与国际协作带动可持续乡建

为了支援四川地震灾区重建，促进当地社区建设和发展，毕马威中国携手中国儿童少年基金会和成都市妇女联合会在四川省彭州市磁峰镇乡村兴建了一座绿色建筑——"毕马威安康社区中心"。

本案建筑师秉承以人为本、节能环保和可持续发展的绿色乡村建设理念，致力于应用节能环保的设计和本地区可更新的素材，包括全面地应用预制的产自本地乡镇企业和永续林区的复合竹结构、竹外围护板和内外竹地板、零污染的农业秸秆材保温墙体、再生集成木材双层保温窗等的建筑材料。

这个低预算的慈善项目得到了多方面志工、物质与资源的大力支持，来自约30个国际和国内的先进绿色企业、科研单位和政府机构，通过合作伙伴的方式提供广泛支持。

面积达450平方米的社区中心于2010年5月17日正式落成并投入使用。目前正用于本地孩童的课余活动和村民的假期培训。本项目以企业赞助、公众参与以及公私合作的国际协作方式和可持续发展的绿色乡村建设理念，推动着乡村社区的发展和建设，具有促进乡村社区教育、文化和娱乐活动发展的研究示范意义。

建筑体现了一系列乡村绿色与智能策略，包括：
- 项目计划支持了公司、个人以及慈善机构在中国的贫困地区通过资金、服务和义工等方面的工作，以赋予想象力的方式落实社会、环保与社区计划
- 采用被动式生态环境调节系统，提升室内环境质量，落实低碳计划

新月形优美的单层曲面结构，有效利用不同时段与季节的太阳和气流，使得多余的冷暖和照明能耗变得没有必要。具体被动设计措施包括：
- 建筑坐落最佳的东南朝向
- 夏季屋顶南侧挑檐遮挡夏日；冬季南侧高窗引入温暖的阳光
- 屋顶太阳能导管引入自然采光，平衡室内自然光分布
- 南侧景观竹杆花架遮阴
- 北立面带状减窗减少冬季热损失
- 曲面屋顶顺导盛行北风
- 通过穿堂风和高窗烟囱效应自然通风
- 建筑底层架空促进自然通风并减少基地影响
- 良好保温性能的空心墙
- 双层节能木窗

创新地广泛应用新型环保低碳建材：
- 全面地应用预制的产自本地乡镇企业和永续林区的复合竹结构
- 研究表明，因为固碳的因素，本项目的竹材基本上是碳中和
- 无污染的农业秸秆纤维板材保温墙体和屋顶构造
- 复合竹材建筑围护和室内外地板
- 复合集成木环保窗

其他的可持续特点：
- 先进乡村教育智能通讯与教育设施，包括CCTF乡村安全应急体验中心
- 在设计建设中融合了场地中拥有两千年历史的"土地庙"
- 竹梁柱结构采用榫卯形式提升建筑的抗震性能
- LED节能照明
- 本地的开放式生态农业与乡村绿色生活方式培训
- 景观融合本地物种和用于乡村灌溉的雨水利用

8. Teenagers' training classroom
9. The roof solar catheter into natural lighting
10. Double energy-saving wooden window
8. 儿童少年安全应急培训教室
9. 屋顶太阳能导管引入自然采光
10. 双层节能木窗

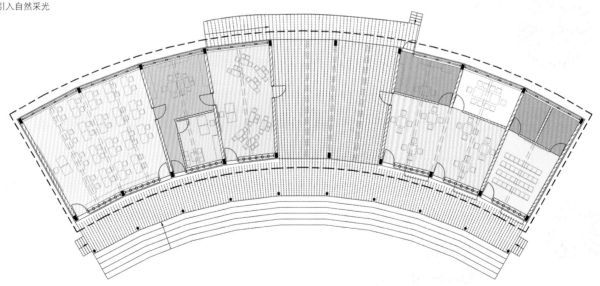

Community Centre Floor Plan: 社区中心平面图：
Pubilc space — 公共空间
Teenagers' training classroom — 儿童少年安全应急培训教室
Children's room — 安亲班
Library — 图书馆
Teacher's office — 教师办公室
Entertainment room & tea house — 社区娱乐中心和茶室
Training room — 培训室
Meeting room — 会议室
E&M and logistics — 机电房及后勤

ZHANGJIAWO ELEMENTARY SCHOOL
Xiqing District, Tianjin
Vector Architects + CCDI

天津西青区张家窝镇小学

天津 西青区
直向建筑+中建国际

Gross Floor Area: 18,000m²
Design/Completion Time: 2008-2009/2010
Architect: Vector Architects + CCDI
Design Architect: DONG Gong
Partners of Vector Architects: DONG Gong, Chien-ho HSU
Collaborating Architects of CCDI: LV Qiang
Project Architect & Site Architect of Vector Architects: WANG Nan
Structural Mechanical Electric Plumbing Engineer: CCDI
Photographer: SHU He
Material: Exterior Stucco, Wood Panel, Wood Louver, Perforated Metal Panel
Structure: Concrete Frame, Steel Truss

建筑面积：18000平方米
设计/建成时间：2008-2009年/2010年
建筑单位：直向建筑+中建国际
建筑设计：董功
直向建筑合伙人：董功，徐千禾
中建国际合作建筑师：吕强
项目建筑师、场地建筑师：王楠
结构及水暖电工程师：中建国际
摄影师：舒赫
运用材料：室外涂料、木板、木格栅、穿孔金属板
结构构造：混凝土框架、局部钢桁架

The goal is to establish a unique place within the school that encourages interaction between the students and teachers through their daily learning and teaching life. The basic programme consists of 48 classrooms, a number of special programme classrooms, cafeteria, training gymnasium, administration areas and an outdoor exercise field.

The design process starts with an analytical research of the spatial pattern of interactive activities, both in plan and in section. A series of physical study models were built along the process, in order to seek the most reasonable spatial and programmatic layout. Eventually the best location of the primary interactive space is discovered to be on the first floor, sandwiched by regular classroom floors, and connected to the skylight through the central atrium, where natural ventilation was maximised. The space is defined by the surrounding special programme classrooms, and extends itself to a green roof deck at the south side, which is also the pivot point of the site arrangement. The deck connects to the main school entrance, the outdoor fields, and different parts of the building at different heights by stairs, ramps and bridges.

Such a "platform", consisting of indoor space and outdoor deck, not only generates and amplifies energy of interactions, but also adds visual characters to the exterior building appearance because of the application of distinctive materials and space modules.

A series of green technologies are proposed in this project, such as geothermal system, storm water management, green roof, permeable landscape, passive ventilation, maximised natural daylight and recycled material.

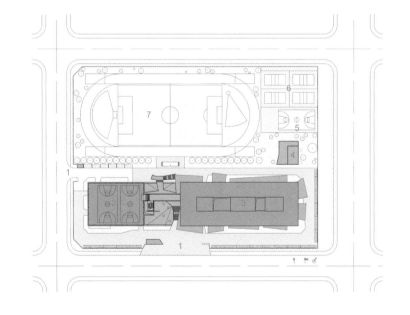

Site Plan:
1. Site entrance
2. Rooftop basketball court
3. School building
4. Toilet
5. Outdoor basketball court
6. Outdoor volleyball court
7. 300m standard running track

总平面图：
1. 场地入口
2. 屋顶篮球场
3. 教学楼
4. 洗手间
5. 露天篮球场
6. 露天排球场
7. 300米标准跑道

我们希望在这个小学设计中着眼于对于"教"与"学"这种生活方式对于空间的需求,尝试提供学生和老师、学生和学生之间充分而富有层次的交流的机会和场所, 在我们看来,这是当前国内的教育建筑的模式化设计中所缺失的要点。

小学的规模为48个班,主要包括普通教室,专业功能教室, 食堂,风雨操场,办公室,室外活动场地。设计起始于对交流空间的行为和空间模式的研究和分析。为了寻求最合理的空间功能布局,我们在过程中进行了一系列手工模型研究。最终,我们将一个共享的交流"平台"设置在二层,它像三明治一样被一层和三四层的普通教室夹在中间,最大程度上带来该

Elevation 立面图

1. View of entrance
2. Northwest view
1. 入口
2. 西北方向全景

空间使用的易达性和必达性。而各个年级交叉,教学形式相对自由,师生交流互动最为频繁的专业功能教室则成为这个交流"平台"的功能载体。

整个建筑活力最强、能量最集中的空间通过一个中庭在顶部获取自然光和加强自然通风,同时延伸出室外,和位于南侧的一层绿色屋顶平台相通,成为连接建筑各部分和教学楼前后景观的中心枢纽。由于功能的特殊性带来的立面材料和开间节奏的特殊性,构成建筑鲜明的室外视觉特征。

我们在设计中倡导运用一系列的绿色环保措施,主要包括地源热泵、绿色屋顶、可渗透景观、自然通风和采光最大化等。

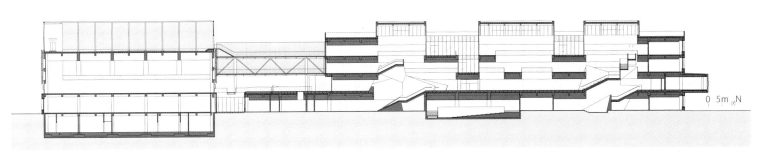

Elevation 立面图

3. Night view of special classrooms
4. Night view of entrance
3. 专业教室
4. 入口夜景

Detail (Right):
1. Metal protective screening
2. H steel column
3. Lesco shading curtain wall keel
4. Lesco shading panel
5. Painted exterior wall
6. Frosted glass
7. Hidden framing glass curtain wall
8. Metal windowsill dagger board
9. White glass
10. Metal grids

墙身详图（右图）：
1. 金属防护网
2. 工字钢柱
3. 绿可木遮阳幕墙龙骨
4. 绿可木遮阳板
5. 涂料外墙
6. 磨砂玻璃
7. 隐框玻璃幕墙
8. 金属窗台披水
9. 白色玻璃
10. 金属箅子

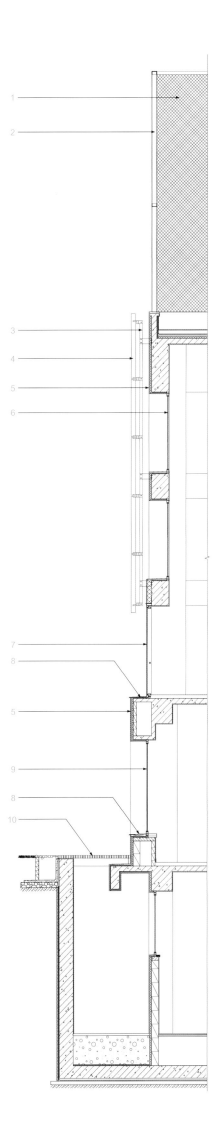

5. View of the gym and the ground floor roof deck
6. View of the ground floor roof deck
5. 风雨操场和一层屋顶平台
6. 一层屋顶平台

Detail:
1. Roof planting
2. Metal sideboard
3. Painted exterior wall
4. White glass
5. Lesco floor

墙身详图：
1. 屋顶植草
2. 金属收边板
3. 涂料外墙
4. 白色玻璃
5. 绿可木板地面

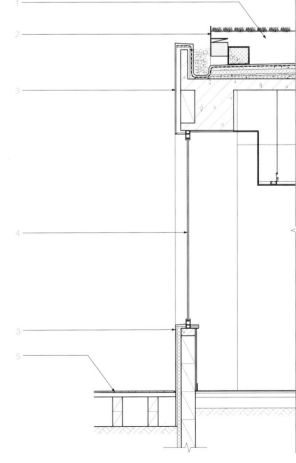

7. View of southeast façade
7. 东南立面

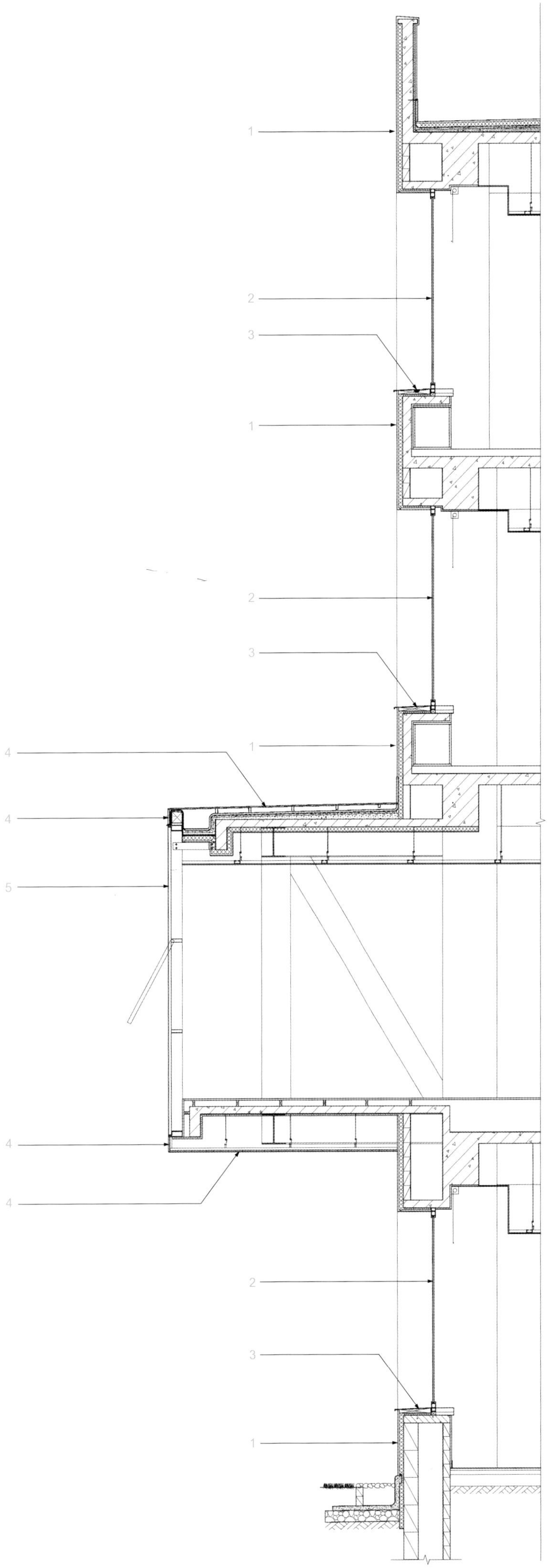

Detail (Right):
1. Painted exterior wall
2. White glass
3. Metal windowsill dagger board
4. Lesco panel
5. Hidden framing glass curtain wall

墙身详图（右图）：
1. 涂料外墙
2. 白色玻璃
3. 金属窗台披水
4. 绿可木板
5. 隐框玻璃幕墙

Eco Energy-Saving Strategy
生态节能策略

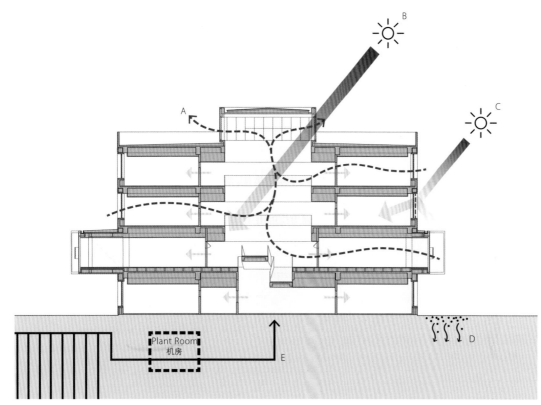

Application of New Energy Technology
E. Ground Source Heat Pump System
The GSHP system uses geothermal energy (coolness or heat)from shallow soil. Through heat pump technology (using Carnot cycle and reverse Carnot cycle to transfer coolness and heat), it transfers heat from the soil to the building in winter and coolness from the soil to the building in summer, forming a annual cycle of coolness and heat. The advantages are efficient, stable, environment-friendly, low maintenance cost, long service life and space-saving.

新型能源技术的应用
E. 地源热泵系统:
利用储藏在浅层土壤中的地能（冷量或者热量），通过热泵技术（利用卡诺循环和逆卡诺循环原理转移冷量和热量的设备）在冬季把热量从地下土壤中转移到建筑物内，夏季再把地下的冷量转移到建筑物内，一个年度形成一个冷热循环。它的优点在于高效节能、稳定可靠、无环境污染、维护费用低、使用寿命长以及节省空间。

Application of Low-Tech Energy-Saving Technology
A. Natural Ventilation and Atrium Draft
In summer, the four-storey atrium and the surrounding openings on the outer wall and internal wall will form a vertical draft effect, which would accelerate interior air flow and lower interior temperature. In winter, the atrium high windows and windows on the surrounding outer walls will be closed, while the extensive glazed walls will introduce sunlight effectively, forming a glasshouse, which will reduce the energy consumption largely.
B. Natural Lighting
All the rooms (except the plant rooms) have large exterior windows, which ensure enough natural lighting even in cloudy days.
C. Outdoor Shading
With extensive openings (in interior gym and part of classroom external windows), in order to avoid side effect of sun radiation to interior environment, the designers use some simple yet effective outdoor shading (outdoor grids).
D. Permeable Paving
The permeable paving will dredge surface runoff effectively, simple and practical.
F. Green Roof
Green roof will beautify the environment, as well as reduce the roof thermal radiation.

低技节能技术的应用:
A. 自然通风及中庭拔风:
在夏季，利用中庭的四层高度，结合周边各房间外墙及内墙上的开窗，形成竖向拔风的工作原理，加速室内空气流动，降低室内温度；在冬季，将中庭高窗及周边房间外墙窗户关闭，中庭大面积的玻璃幕墙能有效将阳光引入，因此形成玻璃温室，减小对电能的消耗。
B. 自然采光:
所有房间（除机房）均设有足够大的外窗，满足即使在阴天的条件下，获得足够的使用光线。
C. 室外遮阳:
在大面积的开窗情况下（风雨操场及部分教室外窗），为避免太阳辐射对室内环境产生的副作用，采用简单有效的外遮阳方式（室外格栅）。
D. 场地透水地面:
有效疏导地表雨水，简单实用。
F. 屋顶绿化:
美化环境，降低屋顶热辐射。

8. Corner view
8. 转角局部

9-11. View of interior atrium
9–11. 室内中庭

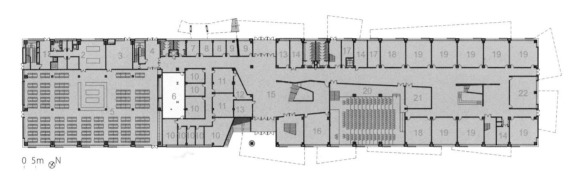

Ground Floor Plan:
1. Passage way
2. Serving area
3. Fire safety
4. Passage way
5. Café
6. Terrace
7. Psychologist
8. First aid
9. Broadcasting office
10. Executive office
11. Meeting room
12. Storage
13. Security room
14. Plant room
15. Entrance hall
16. Science lab
17. Office
18. Multi-use classroom
19. Standard classroom
20. Multi-function hall
21. Equipment storage
22. Exhibition area

一层平面图：
1. 走廊过道
2. 服务区
3. 消防安全教室
4. 走廊过道
5. 咖啡厅
6. 平台
7. 心理咨询室
8. 急救室
9. 广播室
10. 行政办公室
11. 会议室
12. 仓库
13. 保安室
14. 机械设备间
15. 入口大厅
16. 科学实验室
17. 办公室
18. 多功能教室
19. 标准教室
20. 多功能厅
21. 设备仓库
22. 展览区

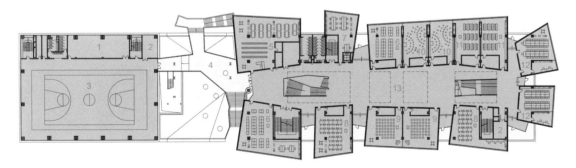

First Floor Plan:
1. Sport equipment room
2. Plant room
3. Indoor gym
4. Passage over 1/floor roof
5. Library/reading room
6. Calligraphy room
7. Office
8. Craft classroom
9. Music classroom
10. Art classroom
11. Science classroom
12. Computer lab
13. Mingling room

二层平面图：
1. 体育器械室
2. 机械设备间
3. 室内健身房
4. 一楼顶端走廊
5. 图书馆/阅览室
6. 书法教室
7. 办公室
8. 手工教室
9. 音乐教室
10. 艺术教室
11. 科学教室
12. 电脑实验室
13. 交际室

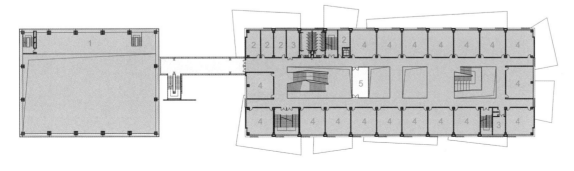

Second Floor Plan:
1. Temporary stands
2. Office
3. Plant room
4. Standard room
5. Terrace

三层平面图：
1. 临时支架
2. 办公室
3. 机械设备间
4. 标准教室
5. 平台

SWIMMING HALL AT TAIWAN BUSINESSMEN'S DONGGUAN SCHOOL (TBDS)
Huangyong, Dongguan, Guangdong
Wang Weijen Architecture

水合院——东莞台商子弟学校游泳馆
广东省 东莞市 潢涌
王维仁建筑设计研究室

Site Area: 1,751m²
Gross Floor Area: 3,000m²
Completion Time: 2009
Architect: Wang Weijen Architecture
Associate Designer: South China Institute of Architectural Design
Photographer: Wang Weijen Architecture
Client: Taiwan Businessmen's Dongguan School

占地面积：1751平方米
建筑面积：3000平方米
建成时间：2009年
建筑设计：王维仁建筑设计研究室
合作建筑师：华南建筑设计研究院
摄影师：王维仁建筑设计研究室
业主：东莞台商子弟学校

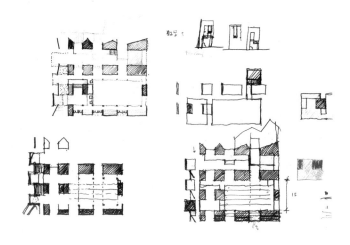

The Swimming Hall at TBDS creates sequential water courtyards with spatial hierarchy and rhythm of solid and void. The project challenges the conventional long-span design for sport facilities which is usually found in large volumes. It re-programmes the swimming activities into several different sizes of pool: L, M, S, and XS. By adopting various scales of pool for different age groups and different ways of "swimming", it cultivates student's senses for appreciating the culture of water.

The integrated settings enclose the campus plaza and open up to it with a series of slopes, decks and colonnades of different scales. The building intends to be a facility that is able to "breath": be able to open itself up completely during the summer, allowing a sufficient ventilation, and close off during the winter to become a small-scale interior swimming hall, with sunshine to reduce loss of heat. The array of colonnades and the contrast of solid and void make the building form a strong hold. On one hand, the slanted roof can be used as a base of solar panels; on the other hand, it represents the folds and changes of traditional architecture.

Beyond the PE Classes
The design originates from TBDS decision to build an interior swimming hall at the end of the campus, on the last piece of remained field between the school buildings and greens. With the school's open mind and supports, the architecture design challenges the conventional long-span design for sport facilities.

Moving Sizes: L, M, S, XS
The first step of design is to re-programme the architecture: how to make students from 6 years old to 18 years old find suitable space scales. Just like L, M, S, and XS sizes of clothes, according to different age groups and phases of swimming learning, the designers arrange swimming pools of L, M, S, XS, children paddling pool, SPA and lotus pond. The design hopes that the space definition of different sizes is in a flowing relationship, not only in visual sense, but also a continuation in lines and functions.

Space of Water Courtyards
The second step is to construct basic spatial relationships: it uses 19 stone walls of different scales to arrange a colonnade space. The heavy solid walls and the void spaces define assembly water yards of different sizes. Not only the heavy walls and light water contrast each other in visual and texture, the logic of colonnade also reduces the structural span, therefore reducing the cost indirectly. The high windows formed by slanted concrete plates introduce direct and indirect light to the water, incorporating swimming into the rhythm of architecture. The water courtyards are reminiscent of Rome bathing pool: same as bathing pools, swimming pools also have their function and culture.

Continuous Landscape
The third step is to make interaction between architecture and human landscape. The architecture not only frames the trees and students running on the playground in water's reflection, its high windows open the roof to

North Elevation 北立面图

South Elevation 南立面图

blue sky. Firstly, the grass slope facing the basketball court on the south is an extension of stone walls; secondly, it is a natural staircase to the activity plaza. The water courtyards on the south of the swimming hall become a broad semi-outdoor veranda. They combine the children paddling pool and the activity plaza and imply the restaurant and dormitory on the east.

A Breathing Building
The paradox of ordinary interior swimming pool is the heating assumption in winter and cooling assumption in summer, so the fourth step of design is to consider how to reduce the warm pools' size in winter and expand the swimming pools in summer. In that way, the energy assumption will be reduced largely. The architects believe that architecture should form a continuous space in the campus.

东莞台商学校游泳馆意图创造一个有不同空间序列层级与尺度的水合院空间文化，打破常规的大跨距、大量体体育建筑的设计模式。设计将原有八水道比赛型制式的功能任务打散成大、中、小不同尺度的泳池，与室内、半室外和室外的不同空间。更重要的是，建筑透过空间虚实的对比与层次序列的趣味安排，使学童充分享受水的空间文化与乐趣。

建筑的整体配置围合了校园的大活动广场，以不同尺度大小以斜坡、平台与柱廊向广场开放。建筑本身意图成为一个能呼吸的有机体：大面积的开窗与天窗在夏季完全开敞，充分的对流空气；冬季则关闭门窗成为小量体的室内温水池馆，引入阳光并减少热能的流失。建筑造型有如高低两组的大型柱列依层级排列，虚实对比；微倾斜的屋顶板一方面作为太阳能吸热板的底座，一方面展现传统建筑的起折变化。

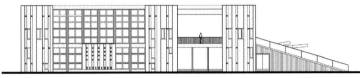
West Elevation 西立面图

在体育课之外

设计起源于东莞台商学校决定在校园终端、整齐的方院校舍和大树草地的最后一片空地,盖一栋室内的游泳馆。在学校的开放和支持下,建筑的设计挑战了原来大结构跨距的体育建筑的样板造型。

移动的尺码:L, M, S, XS

设计第一步,是建筑再计划的过程(Re-Programming):思考如何让6岁到18岁的学生们都在这里找到自己适合的空间尺度。如同穿衣服的大中小尺码,设计根据不同年龄层的学生和游泳学习的不同阶段,安排了大中小尺度和深度的泳池和幼儿的戏水池,还有水疗池和景观的莲花池。设计同时希望大、小水池的空间界定是一种流动的关系,不只是视觉上,也是动线和功能上的连续。

水院的空间

设计第二步是建筑基本空间的关系:利用19块大小不同的石墙量体排列成疏密松紧的柱列空间,由厚重的墙实体和其所界定出的虚体空间组合成大小不同的序列水院。墙体与轻质的水体形成了视觉和触觉上的对比,柱列的逻辑也减少了结构跨距。屋顶利用混凝土斜板形成的高窗,引入打在水面上的直接和间接的阳光,将游泳的运动融入了建筑的韵律。水院空间让人联想到罗马公共浴池场景:泳池如同浴池,是功能,也是文化。

连续的地景

设计第三步,是建筑和自然与人文地景的互动:利用建筑量体框出的,不只是水面倒影延伸出的校园树木和操场上跑步的同学,还有随着高窗的视线和倾斜的屋顶板开展而出的蓝天白云。面对南侧篮球场的大草坡,一方面作为石墙实体的延伸,另一方面作为面向活动广场的自然阶梯。序列的水院在游泳馆南侧,转化成宽广的半室外连廊,结合幼儿的戏水池和南侧的广场活动,在方向上更暗示了向东侧连接的餐厅和宿舍。

呼吸的建筑

一般室内温水游泳池最奇特的地方,在于冬天的加热耗能和夏天温室效应的冷气空调耗能。设计第四步,是思考如何让建筑在冬天的温水池面积和量体缩小,增加日照保温节能;夏天如何让泳池范围扩大,开放对流通风,减少室内直接日照,并连通南侧遮阳柱廊。东莞台商学校的游泳馆没有强烈的视觉形象,因为我们相信建筑更应该是校园里连续的空间。

Diagram:
1. Slanted roof is installed with solar photovoltaic panels
2. Skylight can protect natural lighting in daytime and avoid direct solar radiation in summer
3. Extensive openings and skylights provide sufficient convection current in summer
4. Pools with different scales use different water exchange systems to reduce energy consumption
5. Warm water pool is located in the centre of the building. In winter the closed doors and windows will reduce energy loss

示意图：
1. 倾斜屋面放置太阳能吸热板
2. 天窗白天能保护自然采光，在夏季遮阳避免直射
3. 大面积的开窗与天窗在夏季带来充分的对流空气
4. 不同尺度的泳池以不同的换水管理机制达到节能的效果
5. 温水池放置于建筑中间，冬季开闭门窗引入阳光用来减少热能的流失

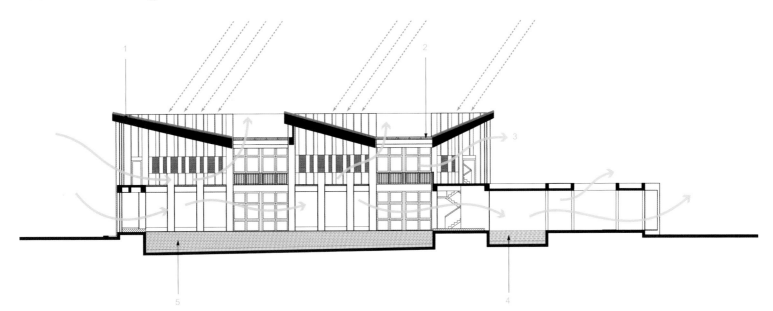

1. Small pool
2. Panorama from southeast corner
3. Grass slope
4. Sun shade
5. Details
6. Enframed scenery of small pool
7. Taking photographs from an elevated position
8. Vestibule

1. 小池
2. 东南角俯拍全景
3. 草坡
4. 遮阳
5. 细部
6. 小池框景
7. 俯拍
8. 前庭

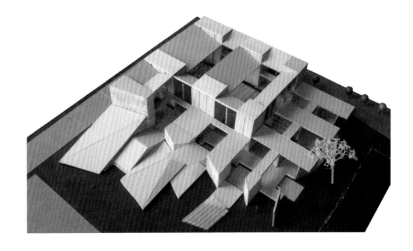

9. West elevation
9. 西侧立面

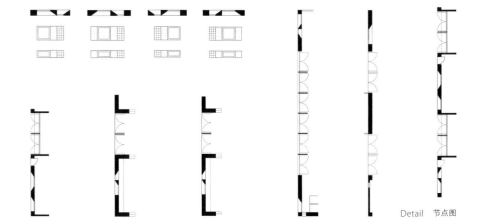

Detail 节点图

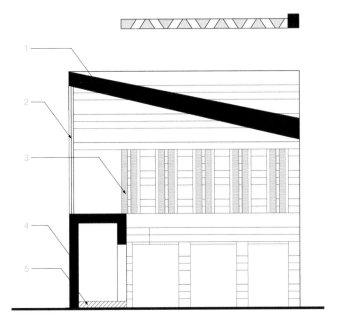
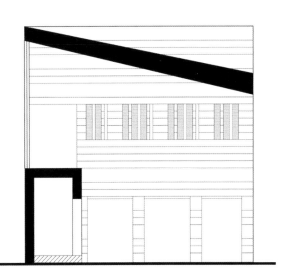

Detail:
1. On-site concrete roof panel, reserved to install solar photovoltaic panels
2. Windows with iron frames
3. Coloured mosaic veneering
4. Pierced block wall, with waterproof finishing
5. Cladding stone

节点详图（上图）：
1. 现浇混凝土屋面板，预留安装太阳能发电板
2. 铁框玻璃窗
3. 彩色马赛克贴面
4. 镂空砌块墙，防水木板铺面
5. 干挂石材

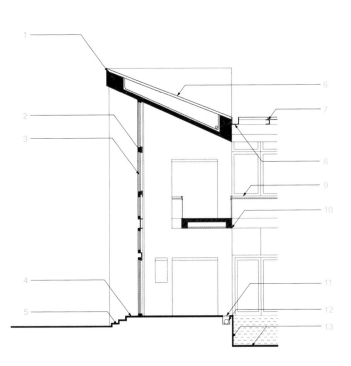
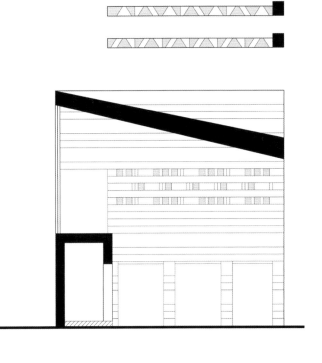

Detail:
1. Concrete beam, with transparent protection painting
2. Iron window frame
3. Openable glass window
4. Anti-skidding tile
5. Plain concrete external stairs
6. Concrete coverplate, with solar photovoltaic panels on the surface
7. Roof window C-type steel beam
8. C-type steel drainage channel
9. Iron bars
10. Mosaic veneering
11. Metal grids
12. Swimming pool edge drainage channel
13. Mosaic veneering of the pool

节点详图（上图）：
1. 混凝土梁，外涂透明保护漆
2. 铁制窗框
3. 可开启玻璃窗
4. 防滑地砖
5. 素混凝土外台阶
6. 水泥盖板批水，外附太阳能发电板
7. 屋顶天窗C形钢梁
8. C形钢排水槽
9. 铁栏杆
10. 马赛克贴面
11. 金属箅子
12. 泳池池缘排水槽
13. 水池马赛克贴面

10. Gallery and infant pool
11. Intersection of large pool and small pool
10. 廊道及幼儿池
11. 大池小池结合部

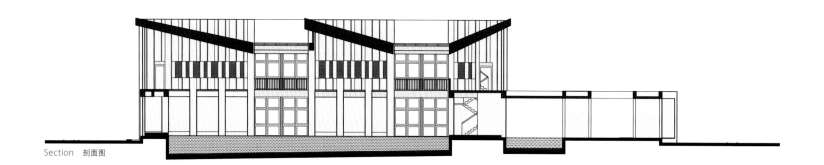

Section　剖面图

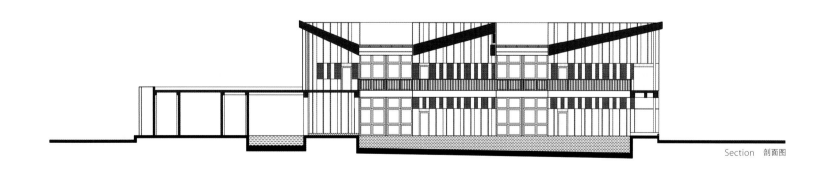

Section 剖面图

12. Large pool
13. Upper lever of large pool
12. 大池
13. 大池上层

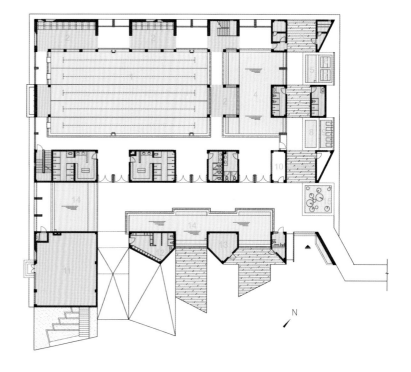

Ground Floor Plan (Right above) and First Floor Plan (Right):
1. Swimming pool
2. Relaxing hall
3. Teacher's area
4. Central pool
5. SPA
6. Hot pool
7. Cold pool
8. Footbath
9. Shower
10. Management room
11. Locker room
12. Changing room
13. Shop
14. Small pool
15. Lotus pond
16. Gym
17. Above the swimming pool
18. Event space
19. Relaxing pavilion

一层平面图（右上图）
二层平面图（右图）：
1. 游泳池
2. 休息厅
3. 教师区
4. 中池
5. SPA
6. 热水池
7. 冷水池
8. 洗脚池
9. 淋浴
10. 管理室
11. 储物间
12. 小更衣室
13. 贩商部
14. 小池
15. 荷花池
16. 健身
17. 游泳池上空
18. 活动
19. 休息亭

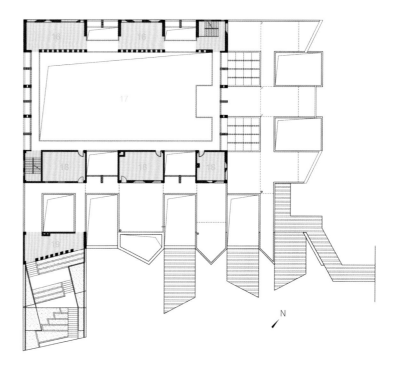

ZED PAVILION
Shanghai
Bill Dunster

上海世博会零碳馆

上海
比尔·邓斯特

Site Area: 900m²
Gross Floor Area: 2,500m²
Completion Time: 2010
Architect: Bill Dunster
Photographer: Emilioo Doiztua

占地面积：900平方米
建筑面积：2500平方米
建成时间：2010年
建筑设计：比尔·邓斯特
摄影师：艾米丽欧·多兹

The ZED pavilion enjoys an excellent position next to the west entrance of the former UBPA 2010 Shanghai World Expo site. This four-storey pavilion provides a 2,500-square-metre exhibition space, demonstrating that a step change reduction in our carbon footprint is possible, at the same time as achieving an increase in the overall quality of life for everyone. Explaining the health, lifestyle and commercial benefits that accompany this approach is one of the pavilion priorities.

The Shanghai Expo ZED pavilion was not only designed to be a zero-carbon building, it also provides the basis for a zero-carbon lifestyle. From food to clothing, transport to consumer goods, leisure activities and work practices, we need to consider the environmental impact of all human endeavours. The series of events, exhibitions and activities held in the pavilion informed and inspired visitors; showing that a zero-carbon lifestyle is possible, enjoyable, fun and rewarding.

ZED pavilion demonstrates a streetscape and two low-cost zero-carbon buildings as a vision of an ordinary mixed-use Chinese street of the future. All of the low-cost innovative building components were sourced in China, and ZED factory has established a supply chain to inform the longer term rollout of zero-carbon urbanism. With China building an area the size of London every year, the ZED factory model of collaboration with local industrial production ensures a successful delivery of workable zero-carbon projects, challenging the current international focus on large, infrastructure-heavy eco city projects, which require excessive up-front investment.

The Urban ZED process pioneered a step change reduction in resource consumption at the same time as offering a higher quality of life for most residents. A conventional environmental approach can offer similar savings in resource consumption, but often asks the public to sacrifice something for the greater good. By taking organic farming principles further, we can re-think agriculture using ZEF methods to create viable and practical solutions.

The results will increase the quality of life for ordinary people and ensure we live in a stable and peaceful society.

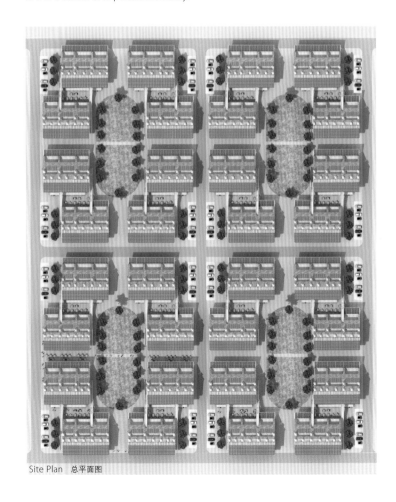

Site Plan　总平面图

148

1. Ventilating cowl and PC solar photovoltaic panels
2. Aerial view
3. Walkway

1. 风帽与PV太阳能光伏板
2. 鸟瞰实景图
3. 连桥实景图

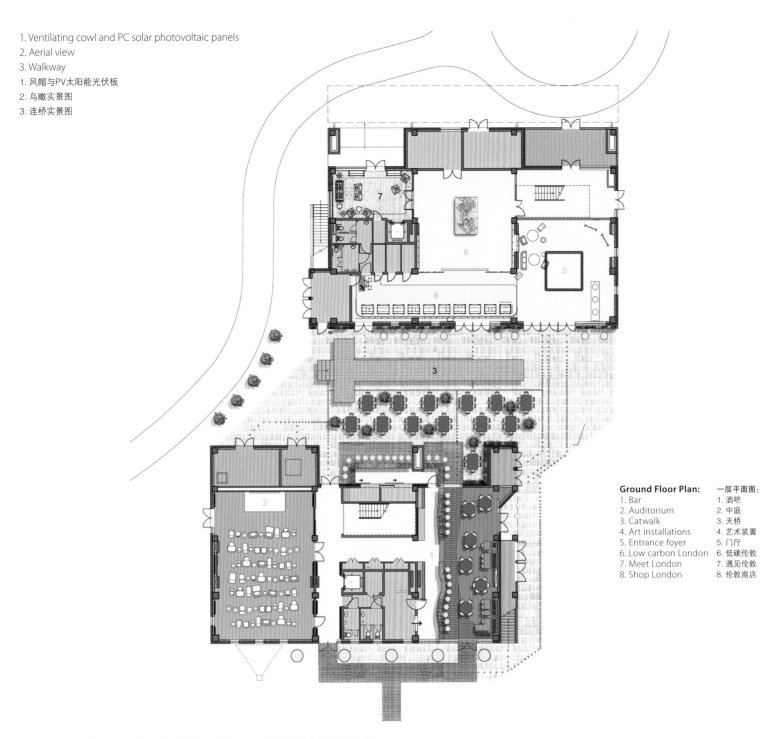

Ground Floor Plan:
1. Bar
2. Auditorium
3. Catwalk
4. Art installations
5. Entrance foyer
6. Low carbon London
7. Meet London
8. Shop London

一层平面图：
1. 酒吧
2. 中庭
3. 天桥
4. 艺术装置
5. 门厅
6. 低碳伦敦
7. 遇见伦敦
8. 伦敦商店

上海世博会零碳馆的位置十分优越，紧邻2010上海世博会城市最佳实践区的西入口。这座四层高的展馆提供了足有2500平方米的展览空间，展示并且证明了目前我们急需缩减碳排放量的必要性，以提高城市居民整体的生活质量。

该展馆的首要任务是阐明伴随着零碳设计而来的健康、生活方式和商业利益方面的效果。从食品到服装、从运输到消耗品、从休闲活动到工作实践，我们需要考量所有人类活动对环境造成的影响。展馆内的系列项目、展览和活动为参观者提供了信息和启发，展示了零碳生活的必要性、趣味性和收益性。

零碳馆由街景和两座低成本零碳建筑组成，展现了未来中国街头的日常景象。所有低成本创意建筑元件都来自中国，而零碳工厂则为零碳城市的长期规划建立了一条生产线。中国每年都将建造一个与伦敦市大小相仿的区域，零碳工厂模型与当地工业生产相结合，保证了零碳项目的多样性，向国际上现有的高启动基金、大规模重生态城市项目发起了挑战。

城市化零碳系统在节约资源的同时还为多数居民创造了更高质量的生活环境，而不同于以往牺牲居民的权益来保存能源。我们可以把有机农业想象成为用一种零碳和可持续发展的方式来耕作。这个结果能提高人民生活水平，使社会更和谐。

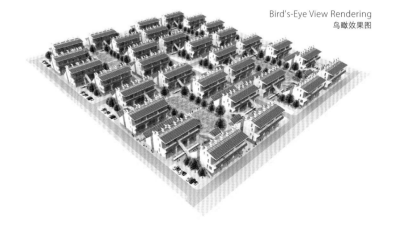

Bird's-Eye View Rendering
鸟瞰效果图

Wind cowls connected to a low-energy dehumidification system provide fresh, cool air without the need for power-hungry air conditioning systems. Residential duplex – south-facing rooms for solar gain and external balcony. PV panels provide electricity; Covered social space between buildings encourages neighbour interactions; A perfect place for a market, café or a place for children to play; All walls, floors and roofs have thermal mass set internally to regulate temperatures and a thick layer of insulation; Exhibition space is double-height – suitable for commercial use – shops and offices, and has cool north light to prevent overheating.

无动力通风系统和太阳能除湿系统缓解了空调的巨大能耗；复式住宅——房间和阳台朝南最大化采光；太阳能光伏板提供电能；栈桥连接建筑，促进邻里互动。市场、咖啡厅和孩童们玩耍的好地方；墙体、地板以及屋顶采用内部控温，并超保温；展示空间采用双倍层高——更适合商业或办公，北向开窗，避免室温过高。

4, 5. ZED pavilion
4、5. 零碳馆外观

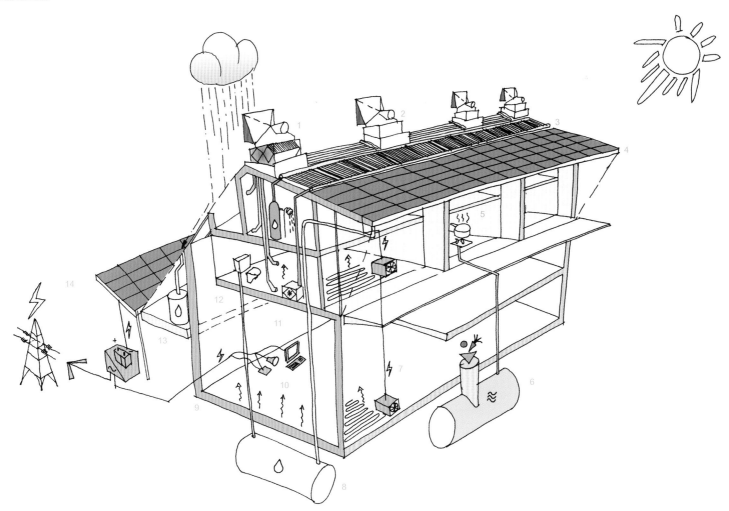

Energy Strategy Diagram (Above):
1. Wind cowl provides wind assisted ventilation with cool recovery
2. Solar hot water used for heating domestic hot water
3. Solar hot water and PV panels
4. Summer sun shaded/warming winter sun used for passive heating
5. Heating provided by occupancy and cooking in winter
6. Food and organic waster placed in biodigester used as gas for cooking
7. Underfloor heating and cooling circuit powered by PV generation
8. Rainwater used in low flush WCs
9. Air tightness line to ensure passive cooling/heating functions when windows are closed
10. PV power is inverted to mains voltage to power lights and appliances
11. Solar thermal unit used to dry dessicant material for passive cooling
12. Concrete in walls and ceilings for passive cooling
13. Rainwater collection for irrigation
14. PV used to charge batteries which can charge from the site wide grid or export to it depending on energy generation
15. Envelope surrounded by super insulation to keep warm in winter and cool in summer

能源战略示意图（上图）：
1. 通风罩利用冷回收提供辅助通风
2. 太阳能热水器用于加热生活热水
3. 太阳能热水器和光电伏板
4. 夏季遮阳/冬季暖阳用于被动式供暖
5. 冬季由居住者和烹饪所产生的热量供暖
6. 食物和有机废料被放入生化池，产生供烹饪所用的燃气
7. 地下供暖和制冷循环系统由太阳能电池板提供能源
8. 雨水用于低水量冲水
9. 气密度线保证了在窗户封闭时的被动式制冷/供暖功能
10. 太阳能通过光电伏板被转换为电源电压，为电灯和电器提供能量
11. 太阳能热单元用于烘干被动制冷的除湿剂
12. 墙壁和天花板上的混凝土保证了被动式制冷
13. 雨水收集用于灌溉
14. 光电伏板为电池充电，电池既能从场地网格中充电，又能将电力输出
15. 由超级保温层包围的外壳保证冬暖夏凉

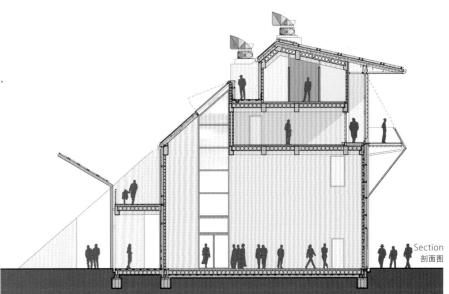

Section
剖面图

151

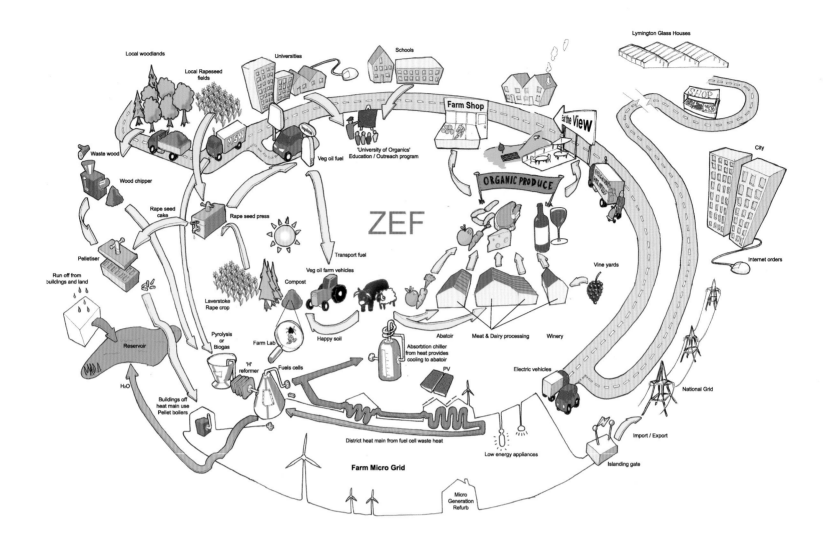

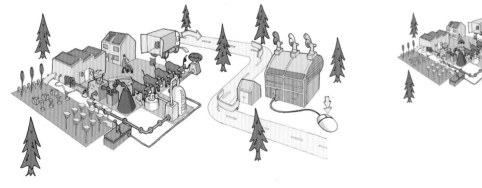
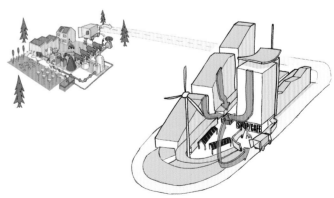

6. Night view
7. Solar photovoltaic panels
8. Block

6. 夜景
7. 太阳能光伏板
8. 街区

With an internal floor area of 111sqm, and an external seating area of over 200sqm, the bar is ideal for lunch, dinner and cocktail parties. The external bar also ties in perfectly with fashion shows held on the catwalk. Internal seating capacity is 62 and the external plaza can accommodate up to 92 guests. The undulating, organically formed suspended ceiling, created from recycled glass bottles, covers the entirety of the bar's ceiling. Lit from above this sculptural piece creates a diffused, mottled light effect in the bar area.

If this room represents the world's people and every chair represents an individual, every individual can have a unique identity, which is created within a fair share of our collective natural resources. These chairs are made from low-carbon or discarded materials. The artist's inventiveness bestows their unique identity, creating an alternative aesthetic to consumer culture and allowing everyone to be different without costing the earth.

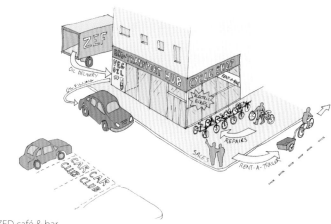

111平方米的空间和超过200平方米的小院，我们创造出了零碳餐厅。无论是用餐或酒会，还是一场时装表演，它都是可容纳92名客人理想之地。一个个空酒瓶被波浪型的吊置在天花板下，覆盖了整个餐厅。灯光从上方射下，营造出斑斓的柔光。这种气氛下年轻艺术家们用他们的旋律让音乐在此回荡。

如果这个空间代表全世界所有人，每一张椅子代表着我们每一个人，那么每个人都有自己的特色，大家公平共享地球上的天然资源。这些椅子都是采用低碳或者废弃的物料做成，艺术家的创造性赋予了它们各不相同的特色，创造一个可替代的美学给消费文化，让每个人都变得与众不同，成为一个没有消耗的地球。

9. ZED café & bar
10. ZED exhibition hall
11. Environmental furniture in ZED designed by students from the Central Academy of Fine Arts

9. 零碳酒吧
10. 零碳馆展厅
11. 中央美院学生为零碳馆设计的环保家具

NANJING ZIDONG INTERNATIONAL INVESTMENT SERVICE CENTRE OFFICE BUILDING
Nanjing Zidong International Creative Park, Zijin Mountain, Nanjing, Jiangsu Province
FU Xiao / IA studio, Institute of Architecture and Urban Planning of Nanjing University

南京紫东国际招商中心办公楼

江苏省 南京市 紫金山东麓 紫东国际创意产业园
傅筱 / 南京大学建筑与城规学院集筑建筑工作室

Gross Floor Area: 2,512m²
Design/Completion Time: 2010/2010
Architect: FU Xiao / IA studio, Institute of Architecture and Urban Planning of Nanjing University
Photographer: YAO Li
建筑面积：2512平方米
设计/建成时间：2010年/2010年
建筑设计：傅筱 / 南京大学建筑与城规学院集筑建筑工作室
摄影师：姚力

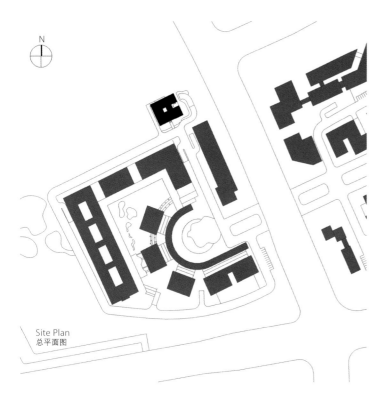

Site Plan
总平面图

Programme
Hall, lounge hall, multi-media presentation room, twelve offices, open office space, two medium conference rooms, a restaurant and a kitchen

Concern
Taking effective measures to save energy

Strategies
1. Adopt PV power generation system to provide indoor landscape lighting and outdoor lighting.
2. Adopt the solar water heater to provide hot water for the washroom, kitchen and bathroom.
3. Put the courtyard in the middle to provide natural ventilation.
4. The ground floor is built on stilts to prevent from damp.
5. GRC integrated external wall insulation systems, prefabricated - modelling, easy to set, less maintenance, reducing the waste of material.
6. Use of ground source heat pump systems to reduce the costs of air conditioning running.
7. Adopt Low-E double course glass, electrical sun-shading louvre and folding curtain.

External Material
GRC integrated external wall insulation systems, Low-E double course glass, sun-shading louvre and folding curtain

Storey
Two floors, the first floor can be changed to the mezzanine by the client.

项目功能
其功能包含门厅、沙盘展示厅、休息厅、多媒体演示厅、12间办公室、一个大空间办公、两个中型会议室及一个小型餐厅和厨房。

项目关注
充分采用各项技术措施，最大减少能耗。

具体措施
1. 采用太阳能光伏电池系统，为部分空间和室外景观提供照明。
2. 采用太阳能热水器为卫生间、淋浴间和厨房提供热水。
3. 采用中心内院布局，为建筑提供自然通风。
4. 建筑底层架空，并与中心内院结合，形成通风循环，防止地面泛潮。
5. 采用GRC集成外墙保温系统，工厂预制，现场安装。采用模数化设计，尽量减少材料损耗。
6. 采用地源热泵空调系统，减低空调运行成本。
7. 采用双层中空Low-E玻璃断桥窗，并采用电动升降遮阳卷帘。

外墙材料
GRC集成外墙保温系统，Low-E双层中空玻璃，遮阳百叶和卷帘。

建筑层数
2层，一层可由业主改造为夹层空间。

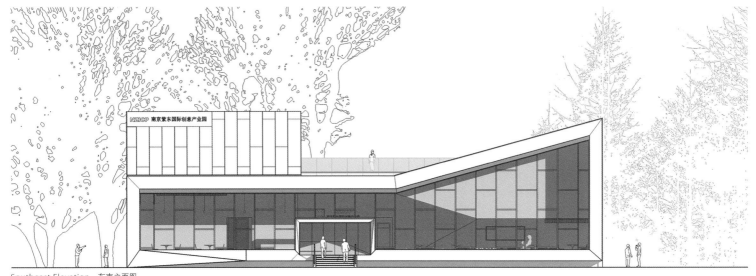

Southeast Elevation　东南立面图

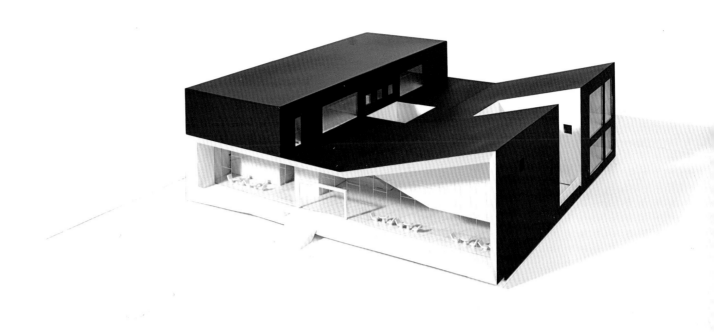

1. The main entrance detail
2. The main entrance
1. 主入口局部
2. 主入口

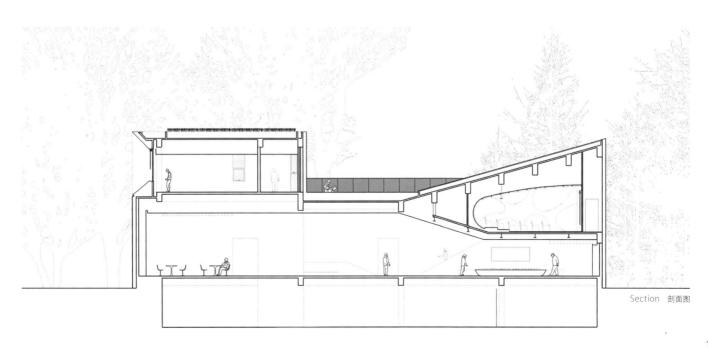

Section 剖面图

3. Southwest perspective
4. Ventilation of inner courtyard
3. 西南透视
4. 通风内院

Space Subdivision
空间划分

Lift
The solar energy PV power panel and solar energy water heater can be placed on the abat-vent
起翘
可在屋顶上放置太阳能光伏电池板和太阳能水泵

Courtyard Implanted
Put the courtyard in the middle to provide natural ventilation and lighting
植入庭院
庭院植入可以让建筑有更多的自然通风和采光

Overhead
The ground floor overhead in order to insure natural ventilation and keep dry
架空
架空建筑使建筑底部保持通风避免一层地面返潮

Concave
The concave shape provides shading for the building
凹入
建筑形体提供自遮阳

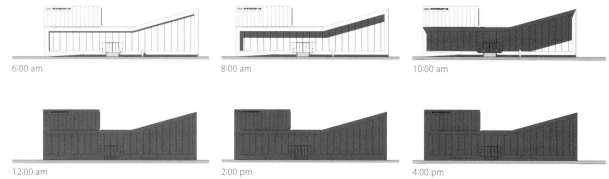

Southeast Elevation
Analysis of Sun-Shade
东南立面遮阳分析图

5. Courtyard
5. 内院

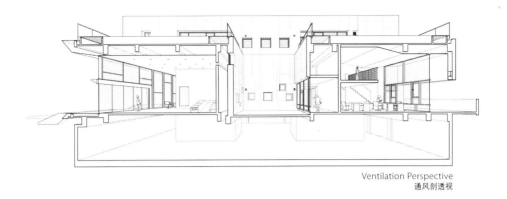

Ventilation Perspective
通风剖透视

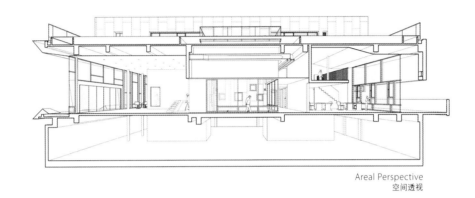

Areal Perspective
空间透视

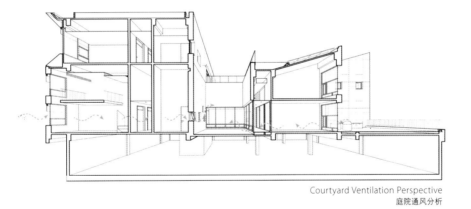

Courtyard Ventilation Perspective
庭院通风分析

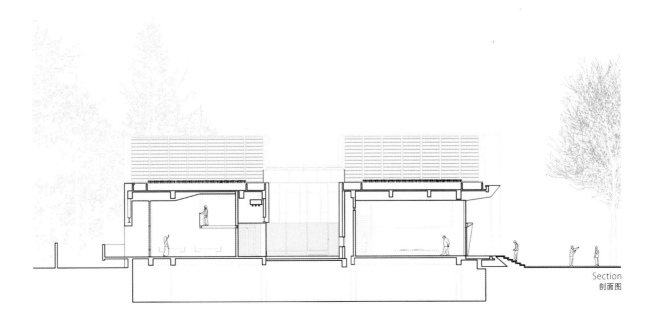

Section
剖面图

163

Ground Floor Plan (Below Left) and First Floor Plan (Below Right):
1. Entrance platform
2. Hall
3. Reception
4. Model showtable
5. Lounge hall
6. Courtyard
7. Office
8. Meeting room
9. Courtyard
10. Cook room
11. Dining room
12. Restroom
13. Bathing box
14. Shower
15. Ventilated caity
16. Air condition well
17. Negotiating room
18. Power device
19. Fire control room
20. Store
21. Lounge
22. Void
23. Planted roof

一层平面图（左下图）和二层平面图（右下图）：
1. 门廊
2. 门厅
3. 总台
4. 沙盘
5. 接待大厅
6. 天井
7. 办公室
8. 会议室
9. 庭院
10. 厨房
11. 餐厅
12. 卫生间
13. 更衣室
14. 淋浴间
15. 通风井
16. 空调井
17. 洽谈室
18. 强电弱电
19. 消防控制室
20. 储藏间
21. 休息室
22. 上空
23. 种植屋面

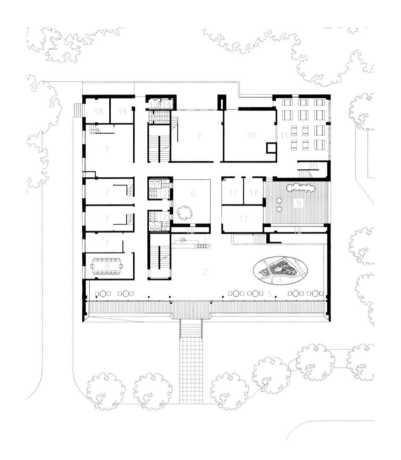

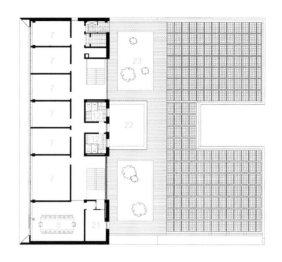

6. Hall
7. View of the roof garden from the stairs
6. 大厅
7. 从楼梯看屋顶花园

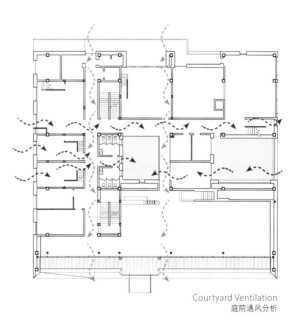

Courtyard Ventilation
庭院通风分析

165

HUAWEI RESEARCH AND DEVELOPMENT PARK
Nanjing, Jiangsu Province
RMJM

华为研发科技园区

江苏省 南京市
RMJM 建筑事务所

Site Area: 279,000m²
Gross Floor Area: 332,000m²
Completion Time: 2010
Architect: RMJM
Photographer: Jason Findley
Client: Huawei Technologies Company Limited

占地面积：279000平方米
建筑面积：332000平方米
建成时间：2010年
建筑设计：RMJM 建筑事务所
摄影师：詹森·芬德利
业主：华为技术有限公司

The Huawei Research and Development Park incorporates offices, laboratories, a data centre and civic plaza and is located to the south of Nanjing, one of the four great ancient capitals of China. The design incorporates a simple, orthogonally arranged architectural composition, with the focus on the sustainable performance and technical aspects of the scheme.

The masterplanning of the site has a synergy with the relationship between the history of Nanjing, the nearby Purple Mountain and the tranquility of the Yangtze River. Integral to the scheme is the integration of the natural typography of the surrounding hills and valleys, with the landscaping being brought through the scheme, blurring the edges between the built and soft environment.

This scheme is a new campus development for China's leading telecoms manufacturer, Huawei Technologies. The project incorporates research office and laboratory accommodation for 10,000 technical staff together with supporting canteen and data centre facilities.

The design proposes a low-rise, orthogonally arranged architectural composition of L-shaped structures to create a series of interlocking courtyards and prioritises harmony with its physical and climatic surroundings.

The building envelopes incorporate extensive solar shading to reduce heat gains and minimise the energy required for cooling purposes. Over 7,000 windows operate through an automated building control system that switches between air conditioning and full natural ventilation modes as external conditions alter. The extensive roof areas provide rainwater harvesting with the water naturally filtered on site through reed beds and providing 100% capacity for all landscape and irrigation requirements year round.

华为研发科技园办公室、实验室、数据中心和城市广场于一体，位于中国四大古都——南京的南部。设计呈现了简单的正交布局建筑群，聚焦于可持续绩效和设计的技术因素。

场地的整体规划与南京的历史、临近的紫金山和宁静的扬子江相契合。设计结合周边山谷的自然形态，让风景融入其中，模糊了建成环境与软环境之间的界限。

本项目是中国顶尖电信制造商——华为技术有限公司的园区设计。项目结合了研究办公室和实验室，可容纳10000名技术员工，并配有食堂和数据中心等辅助设施。

设计采用了低层正交L形造型，建立了一系列连锁的庭院，将重点放在与周边物理环境和气候环境的和谐结合之上。

建筑采用大面积的遮阳板来减少热增量，最小化制冷需求。自动控制系统能控制7,000多扇窗户，使其在空调模式和全自然通风模式之间相互转换。广阔的屋顶区域保证了雨水收集，而雨水在地面上则通过芦苇地自然渗透，保证了全年的景观和灌溉需求。

1. Façade detail
2. Night view
1. 墙面特写
2. 夜景

3. Night view
3. 夜景

Site Plan:
1. Central landscape feature
2. Reception pavilion
3. Civic plaza
4. Canteen facilities
5. Office & laboratory accommodation
6. Training facility
7. Primary entrance gatehouse
8. Secondary entrance gatehouse
9. Demountable office
10. Library
11. Treasury
12. Confidential rooms
13. Data centre

总平面图：
1. 中央景区
2. 接待礼堂
3. 小区广场
4. 食堂
5. 办公室及实验室设施
6. 训练中心
7. 主入口门卫
8. 次入口门卫
9. 多功能办公室
10. 图书数据
11. 接待室
12. 机要室
13. 数据中心

4. Façade
5. Structure
6. Windows
7. Dining room
8. Bridge
9. Corridor
4. 建筑立面
5. 结构设计
6. 窗户细节
7. 食堂
8. 天桥
9. 走廊

ALIBABA HEADQUARTERS
Hangzhou, Zhejiang Province
HASSELL

阿里巴巴新园区

浙江省 杭州市
HASSELL 建筑事务所

Gross Floor Area: 150,000m²
Completion Time: 2009
Floors: Four to seven storeys
Architect: HASSELL
Photographer: Peter Bennetts
Awards: 2011 HKDA Global Design Awards - Bronze Award - Office;
2011 Asia Pacific Property Awards - Highly Commended - Office Architecture China

建筑面积：150000平方米
建成时间：2009年
楼层数：4层到7层不等
建筑设计：HASSELL 建筑事务所
摄影师：彼得·贝内特斯
奖项：2011年香港设计师协会环球设计大奖——铜奖——办公空间；
2011年亚太房地产奖——推荐奖——中国办公建筑

The dynamic campus accommodates approximately 9,000 Alibaba employees and has been designed to reflect the interconnection, diversity and vitality of the company.

The master plan principles for the Headquarters are based on the concepts of connectivity, clarity and community – concepts that are also vital to Alibaba's e-commerce business. These principles guided all design decisions from the single workstation to the greater workplace community.

The campus is arranged around a central open space or "common" surrounded by a cluster of buildings or "neighbourhoods" that vary in height from four to seven storeys.

The built form and the designed "spaces between places" are integrated so that each defines the other. The grand central space is complemented by a series of more intimate gardens that nurture the individual within the larger corporate community. The humanised scale of the built form and the long, narrow floor plates help to create a strong sense of place at a legible scale, and establish physical connection throughout the campus. Visual permeability – or the ability to see into and across the major courtyards into other parts of the complex – is also key to achieving the sense of community and connectivity.

The Hangzhou context has been embraced with garden networks and sunshading screens that represent Chinese ice-pattern window screens which are prominent throughout the city's renowned historical gardens. The sustainable design incorporates features to minimise the campus' environmental impacts while maximising its contribution to the health, wellbeing and productivity of its population.

The flexible, open plan workplace has been designed to be a positive and healthy environment that encourages informal and creative meetings throughout the complex. Hubs, internal and external streets, bridges, roof terraces and strategically placed destination points contribute to the collaborative nature of the workplace. Buildings and floorplates have been arranged and the façade has been designed to maximise access to natural light and air flow to all workstations. Horizontal sunshades on the south façade cut out mid-afternoon sun, reducing the need for cooling. Vertical sunshades cut out western sun and openable windows on opposite façades promote cross ventilation.

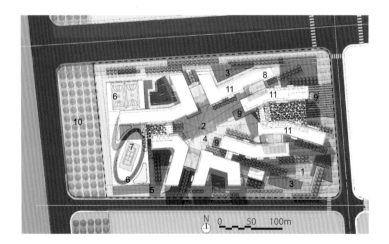

Site Plan:
1. Arrival plaza
2. Ali commons
3. Courtyard
4. Information field
5. Service road
6. Multi-purpose sport court
7. tennis court
8. Running track
9. Structural water body
10. Green belt
11. Roof terrace

总平面图：
1. 入口广场
2. 阿里巴巴公共区
3. 庭院
4. 信息领域
5. 服务通道
6. 多功能运动场
7. 乒乓球场
8. 跑道
9. 建筑水体
10. 绿化带
11. 屋顶平台

Elevation
立面图

Main Eco Features

1. Climatic Characteristics of Its Location:
Located in Hangzhou, Zhejiang, China, which is characterised by a humid subtropical climate with four distinctive seasons, including very hot, humid summers, and chilly, cloudy and dry winters.

2. Atrium Design Strategy:
The built form and the design "spaces between places" are integrated so that each defines the other.

3. Green Materials and Equipments Involved:
The planning and building design promotes passive ESD features, including the use of daylight and shading in the occupied spaces, for energy savings. Low-E glass was also installed for the project.

4. Indoor Physical Environmental Strategy:
Horizontal sunshades on the south façade cut out mid-afternoon sun, reducing the need for cooling. Vertical sunshades cut out western sun and openable windows on opposite façades promote cross ventilation.

5. Other Eco Features:
The heat transmittance factor, solar shading coefficient and the daylight accessibility all comply with the requirements of China's latest energy-saving design standards and bearing in mind the weather conditions in Hangzhou.

Sections
剖面图

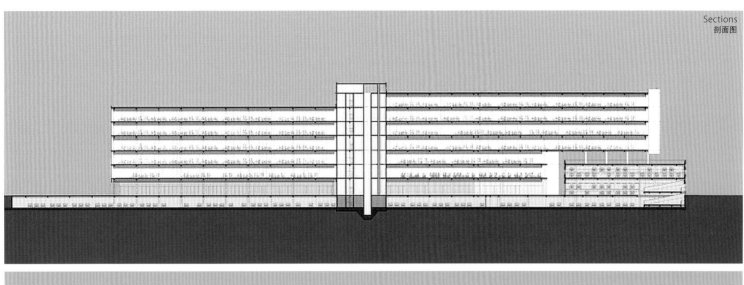

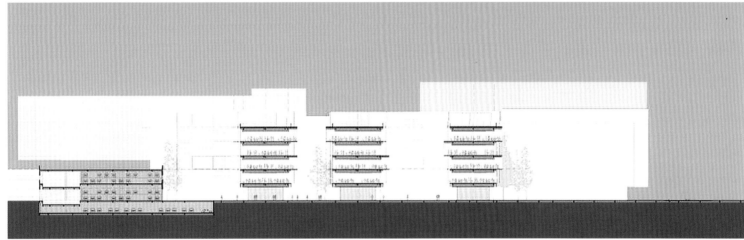

1. Detail of façade
2. Full view of the façade
1. 外墙特写
2. 大厦全景

Sections
剖面图

生机勃勃的阿里巴巴产业园共有将近9000名员工,充分体现了公司的互联化、多样化和生命力。

这栋总部大楼的整体规划以连通性、透彻性和社区性为核心设计理念,这也是阿里巴巴电子商务业务运行的重要理念。这些设计原则贯穿项目设计的始终——从单一的工作台到更大的集体工作区。

整个产业园区围绕中央一个开放区而建,四周大楼簇拥,楼层高度从四层到七层不等。

建筑外形与各建筑间的区域相互融合,彼此界定。巨大的中央广场内装点着一系列花团锦簇的花园,使员工身心愉悦。建筑外形的人性化比例和细长的楼面形成了清晰的轮廓,同时打造了强烈的空间感,并形成了整个产业园的实体连接。这栋建筑综合体各部分间的透视性也是实现建筑社区性和连通性的关键因素。

杭州工业园拥有成网络覆盖的花园和著名的城市历史园林中广泛使用的遮阳篷,遮阳篷代表了中国冰块结构的窗格子。可持续性的设计结合了各方面的优势,将园区的环境影响降至最低,同时最大程度地保证了员工的健康与幸福、提高了其生产效率。

灵活开放的空间布局被设计在一个积极健康的环境中,大厦内随处可召开灵活的创新型会议。城市枢纽、内外街道、桥梁、屋顶平台及所有战略意义的景观都衬托出了工作区互动合作的特点。建筑与楼面都经过整体规划,建筑外立面的设计使建筑拥有最佳光照,并保证各区域内的最佳通风。南立面的水平遮阳板切断了下午的光线,因此无需降温设备。垂直的遮阳板切断了傍晚的光线,对面开关灵活的窗户增加了室内的对流通风。

主要生态特点:
1. 项目所在地的气候特点:
阿里巴巴园区地处中国浙江省杭州市,四季分明,属于潮湿的亚热带气候;夏季湿热,冬季干冷,常是多云的天气。
2. 中庭设计策略:
建筑外形与各建筑间的区域相互融合,彼此界定。
3. 相关绿色材料与设备:
建筑的规划和设计都体现了其被动式静电防护的功能,其中包括建筑空间范围内日光与遮阳篷的利用,从而有效地节约了能源。此外,建筑还安装了低辐射率的玻璃。
4. 室内物理环境的设计策略:
南立面的水平遮阳板切断了下午的光线,因此无需降温设备。垂直的遮阳板切断了傍晚的光线,对面开关灵活的窗户增加了室内的对流通风。
5. 其他生态特点:
该建筑的传热因子、遮阳系数和日光透射率等因素都符合中国最新的节能设计标准,而且建筑设计充分考虑到了杭州当地的气候条件。

3. Activity hub
4. Building elevation
3. 活动中心
4. 建筑立面

Ground Floor Plan: 一层平面图：
General office 普通办公区
Neighbourhood hubs 集散中枢
Destinations 多功能区
Limited access 受控区域
Roof terrace 屋顶花园
Roof (no access) 屋顶（不可进入）

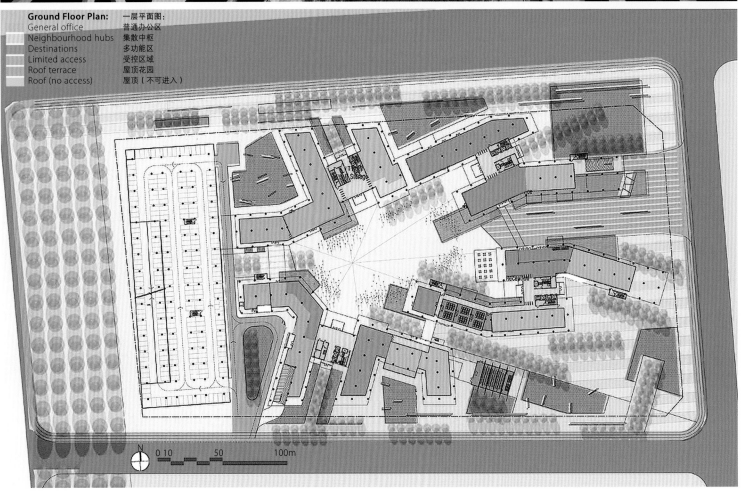

5, 6. Landscape
5,6. 景观

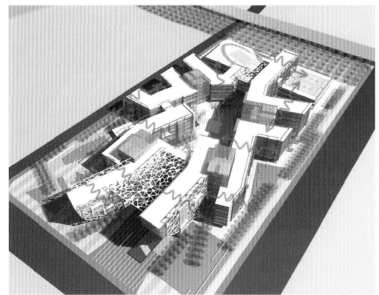
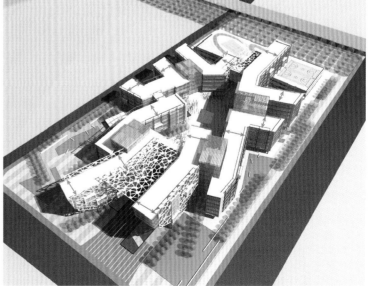

SHANGHAI INTERNATIONAL CRUISE TERMINAL
Shanghai
SPARCH

上海国际港客运中心

上海

思邦建筑设计咨询有限公司

Gross Floor Area: 263,448m²
Completion Time: 2010
Architect: SPARCH
Photographer: Christian Richters, Johnson Xu

建筑面积：263448平方米
建成时间：2010年
建筑设计：思邦建筑设计咨询有限公司
摄影师：克里斯蒂安·里克特斯，约翰逊·徐

The design of the architecture for the cruise terminal site considered the Herculean scale of the cruise ships that will dock alongside. The brief required that 50% of the total construction area be placed underground, including the cruise terminal passenger facilities (planned by Frank Repas Architects), thus freeing up most of the site as a green park terracing down to the water's edge. This development also represents a first in Shanghai for sustainability, incorporating "River Water Cooling Technology"– utilising water from the Huang Pu River as a refrigerant to cool and thereby greatly reduce the buildings' energy consumption during the summer months.

SPARCH's challenge was how to deal with the "under world" as well the architecture rising out of it. Their solution was to create ambiguity as to where the ground plane is, by opening up a honeycomb of sunken courtyards. The buildings appear to disappear into these sculpted holes, providing abundant opportunities to explore connections between the ground and "lower ground" levels. The concept also explored the idea of ripples in the landscape being amplified into standing crystal waves that wrap over the buildings (see concept model). This augmented over time into a second skin that protects the commercial office spaces from their due south orientation, and is populated with semi-outdoor balcony spaces overlooking the Huangpu River. The river front faces the city, and illuminates at night into a herring bone array of delicate curved masts that tie the pavilion buildings together. An intriguing gap appears in the middle – a glazed table top supports amorphous pods on cables. One-, two- and four-storey pods contain cafés, bars and restaurants, hovering over a public performance space below. There is a symbiosis between Shanghai's fun-loving desire for diversity, and SPARCH's approach to design, which has made this architecture a reality.

The quest was to create one "Chorus Line" structure, of an appropriate scale that would sit comfortably beside three 80,000-tonne cruise ships that will dock alongside. This question of scale is also critical in terms of how the scheme is perceived from Shanghai's famous Bund to the south. The scheme needed to have a strong visual presence to be considered a 21st century continuation of the Bund. The response was to wrap the buildings in a fluid steel and glass solar skin, visually tying the accommodation together, and creating a continuous Winter Garden gallery space, which will contain green hanging gardens. The glass façades peel out along the base to shelter a pedestrian route along the newly formed public park.

The lights have just been switched on, exposing a delightful layer of "Amethyst" crystal balconies inside a 400-metre-long herringbone steel and glass skin, clearly visible from the historical Bund to the south.

上海国际港客运中心的建筑设计已经考虑到了今后码头船运的巨大吞吐量，所以在项目任务书中更是要求要将50%的面积作为地下空间，其中包括为客运旅客设计的相关设施，然后敞开所有的场地作为绿色公园，层层叠叠地向黄浦江边推进。此开发项目也是上海可持续发展的先驱作品，融合了"江水冷却技术"将黄浦江的水作为冷却剂，显著降低建筑物在夏季的能耗。

这里SPARCH的挑战是如何处理"地下世界"以及地上的建筑。对此，设计方案是运用向上开放的一系列蜂窝式下沉广场来创造模糊地平面的效果。建筑如同是从这些蜂巢状的洞口中生长出来的一样，从而为人们提供了更多的机会来探索地面和多层地下空间的联系。层层线性起伏的景观设计同样被引伸到建筑的幕墙设计中去。他们为大楼表面提供了第二层幕墙，让朝南的商业办公空间免受强烈光线的影响，同时在两层幕墙之间设置了室外阳台空间以供人们可以俯视黄浦江景。入夜，栓系大楼表面的鲱鱼骨头般曲度精美的排排桅杆被城市的灯光所照亮，并映射在江面上。而在其中两栋大楼之间的空隙处矗立着一个巨大的玻璃"桌子"，桌子里面悬吊着数个不规则造型的吊舱。这些吊舱悬浮在一个公共演出空间之上，位于桌子内部的一、二、四层，功能分别为咖啡馆、酒吧和餐厅。SPARCH的设计和上海的多元娱乐需求融为一体，最终将这一奇特的建筑变为现实。

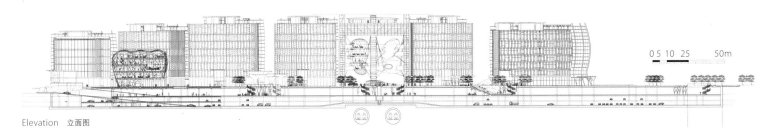

Elevation 立面图

为了创造一个"乐谱"（Chorus Line）式的架构，规模需与将停泊一旁的三艘80,000吨级游轮相得益彰。由于地理位置得天独厚，从上海著名的外滩方向观看本项目，观景效果亦至关重要。该项目需要具备强烈的视觉呈现力，从而成为21世纪上海外滩新的一景。

对策是将建筑群外包一层流线型钢架和太阳能玻璃幕墙，在视觉上连贯各个办公楼，并且建立一条绵延而充满空中绿化的冬季花园空间。玻璃幕墙于地面向外展开，为沿着新建公园绿化带的步道遮风挡雨。夜灯开启，透露出一个个赏心悦目的"宝石"水晶阳台，在400米长的人字形钢架和玻璃壳内，从南面历史悠久的外滩上清晰可见。

1. Entrance
2. Panorama
1. 入口
2. 全景

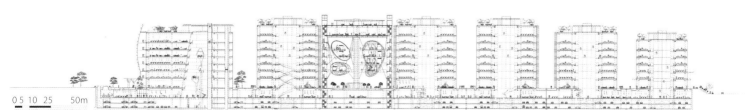

Elevation 立面图

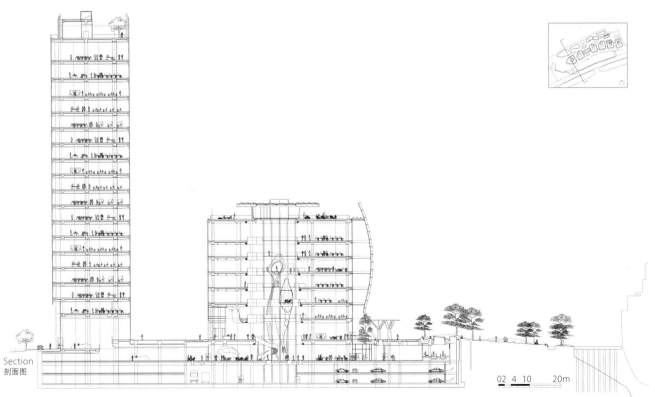

Section
剖面图

3. Gallery building
4, 5. Atrium view
3. 展厅
4、5. 鸟瞰

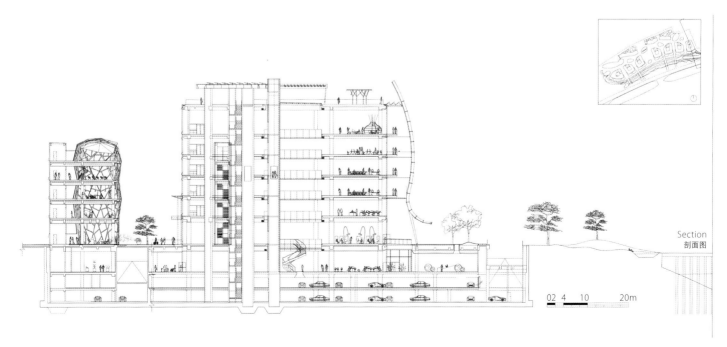

Section
剖面图

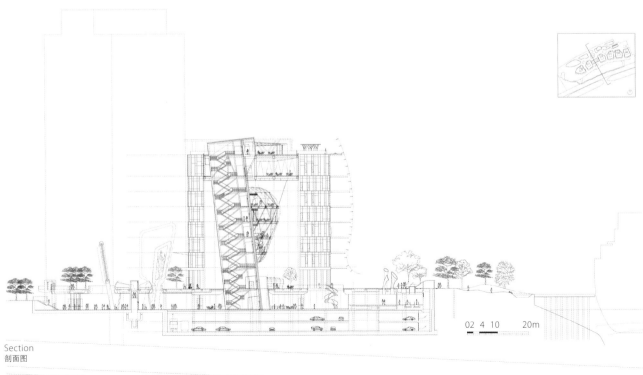

Section
剖面图

Section of Chandelier
吊灯剖面图

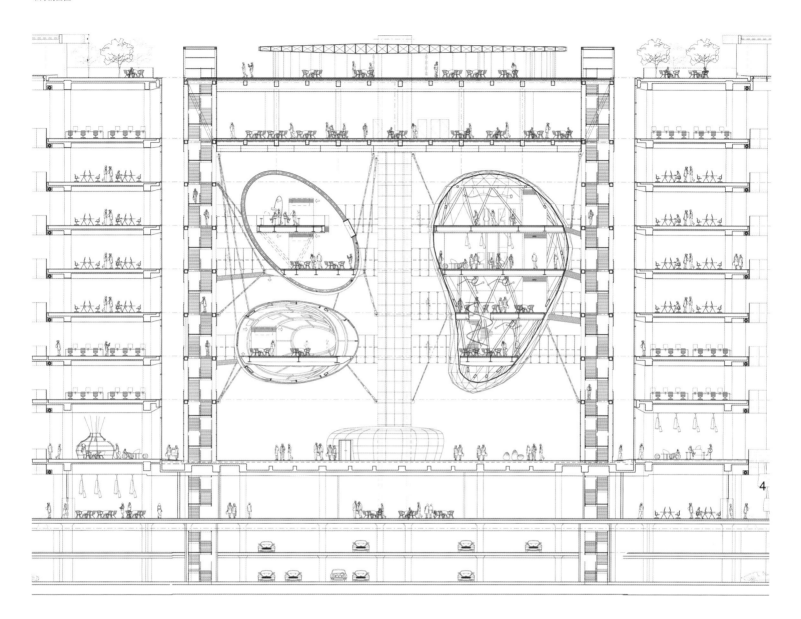

6-8. Chandeliers
6–8. 吊灯

9. Courtyard
10. Façade
9. 庭院
10. 外墙特写

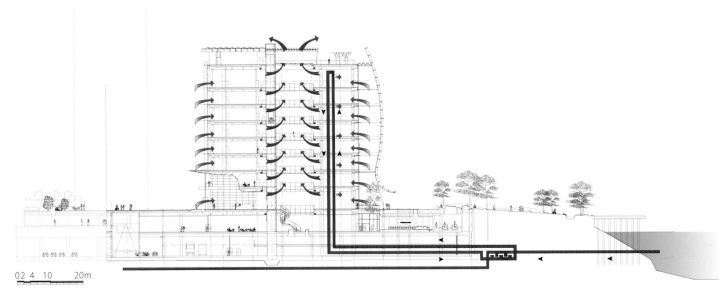

VANKE CENTRE
Shenzhen
Steven Holl Architects

万科中心

深圳
斯蒂文·霍尔建筑事务所

Gross Floor Area: 120,445m²
Public Green Space: 47,288m²
Completion Time: 2009
Architect: Steven Holl Architects
Associate Architects: CCDI
Photographer: SHU He, Steven Holl Architects
Awards : AIA NY Architecture Honour Award, USA, 2010

建筑面积：120445平方米
公共绿地面积：47288平方米
建成时间：2009年
建筑设计：斯蒂文·霍尔建筑事务所
助理建筑师：CCDI
摄影师：舒赫，斯蒂文·霍尔建筑事务所
奖项：2010年美国建筑师协会纽约分会建筑荣誉奖

The building appears as if it were once floating on a higher sea that has now subsided; leaving the structure propped up high on eight legs. The decision to float one large structure right under the 35-metre height limit, instead of several smaller structures each catering to a specific programme, generates the largest possible green space open to the public on the ground level.

As a tropical strategy, the building and the landscape integrate several new sustainable aspects. A micro-climate is created by cooling ponds fed by a greywater system. The building has a green roof with solar panels and uses local materials such as bamboo. The glass façade of the building will be protected against the sun and wind by porous louvres. The building is a Tsunami-proof hovering architecture that creates a porous micro-climate of public open landscape. Simultaneously, the Vanke Centre will be the first, highest rated USGBC, LEED Platinum Certified Project in China.

Renewable Materials

Bamboo – This highly renewable material, which is easily available in China, is used for doors, floors, and furniture throughout the Vanke Headquarters instead of using raw materials or exotic woods.

Green Carpet – InterfaceFLOR Carpet tiles are used throughout the open office area. This carpet is a cradle-to-cradle product, meaning that it is not only produced from recycled materials, but that the manufacturer agrees to collect any damaged carpet and to recycle it into other carpet or products. This carpet contains a GlasBac®RE backing that has an average of 55% total recycled content with a minimum of 18% post-consumer recycled content. It uses recycled vinyl backing from reclaimed carpet tiles and manufacturing waste.

Non-toxic Paint – All paint finishes, as well as the millwork and adhesives are to be low or free of V.O.C (Volatile Organic Compounds) – like phenols and formaldehyde – which can cause various health and environmental problems.

Greenscreen Shading – The Vanke Headquarters uses Greenscreen solar shading fabrics from Nysan – a PVC-free product that contains no V.O.C. (Volatile Organic Compounds). Not only does the product not "off-gas" during its life time, but it is also easier and quicker to recycle and divert to landfills.

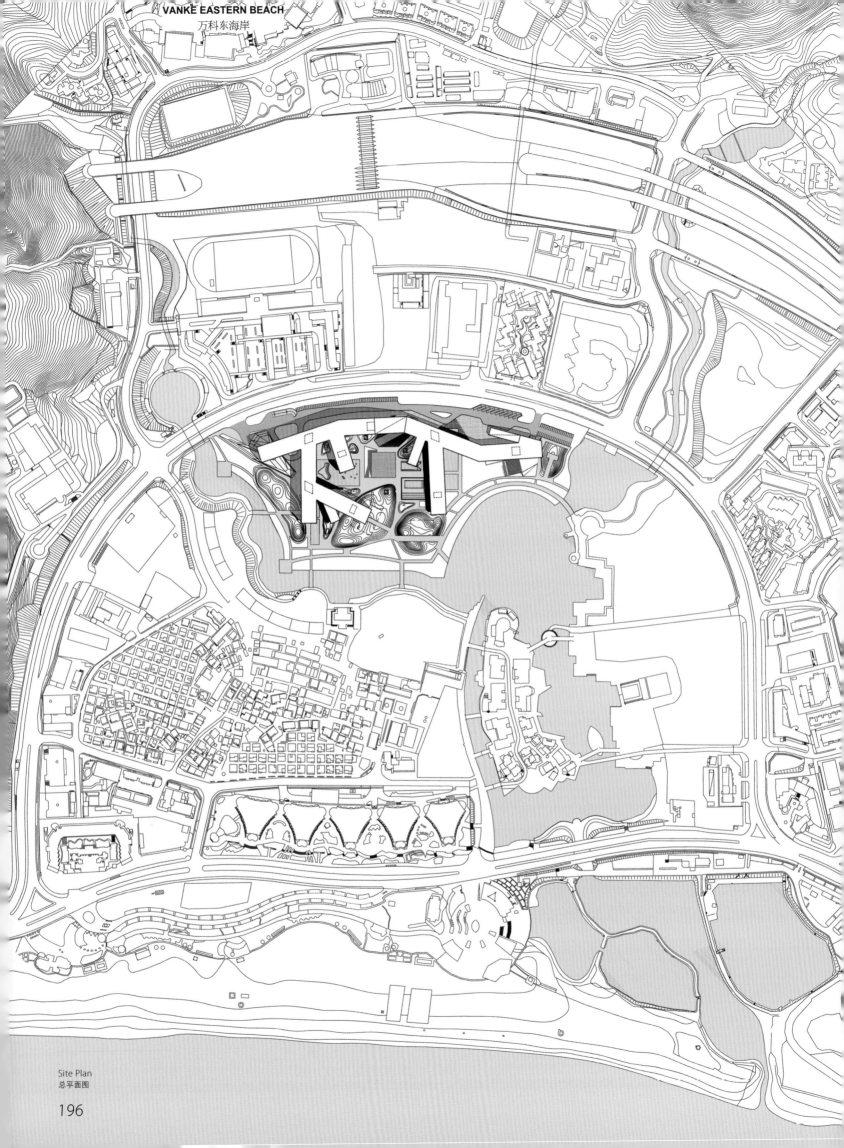

Site Plan
总平面图

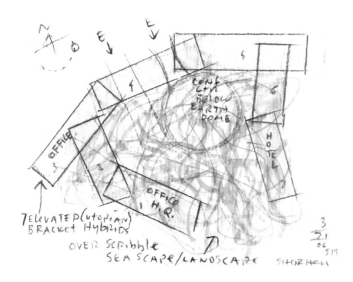
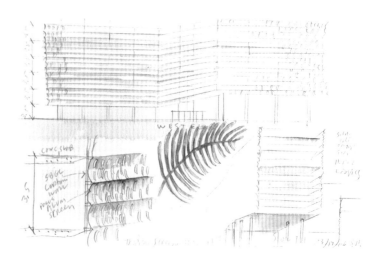

Sustainable Site

The building is sited on reclaimed/stabilised land that forms part of the municipal storm water management system. The lagoon functions as bio-swale/retention pond connected to several adjacent creeks. Part of the landscape architecture water edge proposal designed by Steven Holl Architects is the redesign the municipal landscape bulkhead into a soft-edge planted estuary. As a restorative ecology, the Vanke Centre landscape works to maintain native ecosystems to minimise run-off, erosion and environmental damage associated with conventional modes of development.

The project is both a building and a landscape, a delicate intertwining of sophisticated engineering and the natural environment. By raising the building off of the ground plane, an open, publicly accessible park creates new social space in an otherwise closed and privatised community.

The site area is approximately 60,000 square metres: of which 45,000 square metres is planted. With the addition of the planted roof area of the main building (approximately 15,000 square metres) – the total planted area of the project is roughly equal to the site before development.

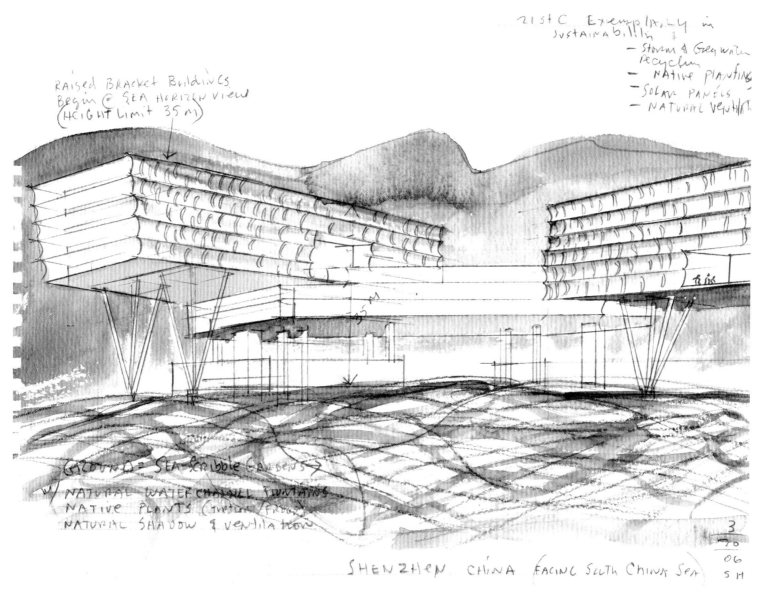

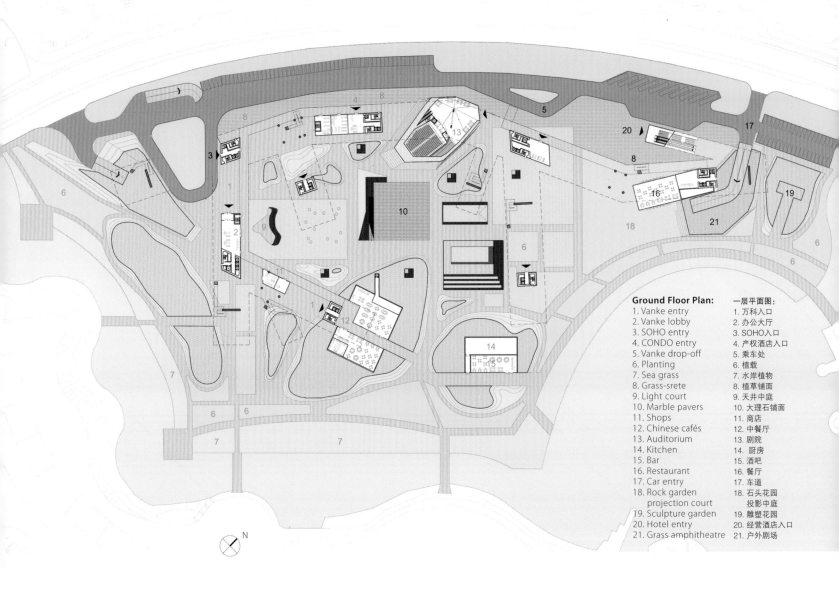

Ground Floor Plan: 一层平面图：
1. Vanke entry　　　　1. 万科入口
2. Vanke lobby　　　　2. 办公大厅
3. SOHO entry　　　　3. SOHO入口
4. CONDO entry　　　 4. 产权酒店入口
5. Vanke drop-off　　　5. 乘车处
6. Planting　　　　　　6. 植载
7. Sea grass　　　　　 7. 水岸植物
8. Grass-srete　　　　 8. 植草铺面
9. Light court　　　　　9. 天井中庭
10. Marble pavers　　　10. 大理石铺面
11. Shops　　　　　　 11. 商店
12. Chinese cafés　　　12. 中餐厅
13. Auditorium　　　　 13. 剧院
14. Kitchen　　　　　　14. 厨房
15. Bar　　　　　　　　15. 酒吧
16. Restaurant　　　　 16. 餐厅
17. Car entry　　　　　17. 车道
18. Rock garden　　　 18. 石头花园
　　projection court　　　投影中庭
19. Sculpture garden　　19. 雕塑花园
20. Hotel entry　　　　 20. 经营酒店入口
21. Grass amphitheatre　21. 户外剧场

1. Elevation detail
2. Overview under construction
1. 外立面细节
2. 施工现场

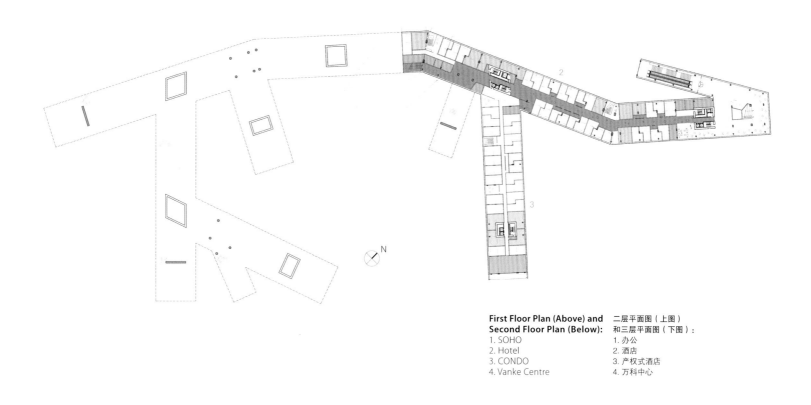

First Floor Plan (Above) and Second Floor Plan (Below):
1. SOHO
2. Hotel
3. CONDO
4. Vanke Centre

二层平面图（上图）和三层平面图（下图）：
1. 办公
2. 酒店
3. 产权式酒店
4. 万科中心

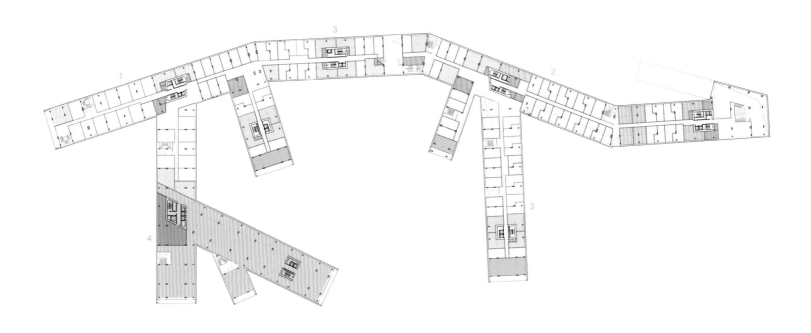

3. Elevation
4. Landscape of outdoor CONDO lobby
3. 外立面
4. 公寓大堂室外景观

建筑看起来就像浮在一片退潮的海水之上，以八根底柱支撑着主要结构。建筑师将35米高（包括底座）的大型结构悬浮在空中，而不是分解成若干个小型结构。这一设计为公众在地面提供了最大面积的绿化空间。

建筑和景观在设计中采用了一些全新的可持续措施。由灰水系统支持的凉水池为项目打造了一个微环境。建筑的绿色屋顶上配有太阳能光电板，并且采用了竹子等本地材料。建筑的玻璃外立面通过多孔百叶窗进行遮阳和防风。建筑具有防海啸性能，在开阔的公共景观中打造了一个可渗透的微环境。同时，万科中心将是中国第一座经美国绿色建筑委员会认证的白金级绿色建筑。

可再生材料

竹子——竹子材料具有高度可再生性能，在中国极易获得，被应用在万科总部的门、地面和家具中，替代了原材料和外来木材。

绿色地毯——整个开放式办公区都采用了英特飞块式地毯。这种地毯是"从摇篮到摇篮"产品，这意味着产品不仅由回收材料制作，而且制造商还同意回收损坏的地毯并将其回收利用于其他地毯或产品中。这种地毯所含的GlasBac®RE衬垫内有55%的回收含量，消费后回收含量最小值为18%。它所采用回收乙烯基衬垫来自回收的块式地毯和工业废料。

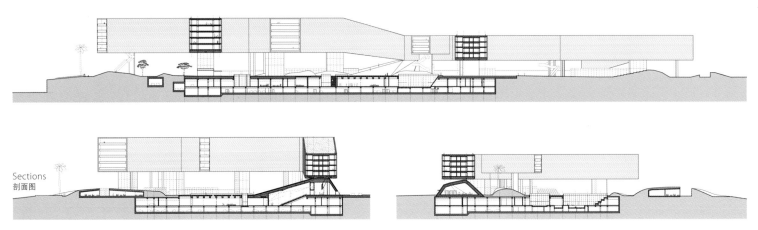

Sections
剖面图

4

无毒油漆——所有油漆材料、木材产品和黏合剂都为低挥发性或无挥发性——没有多酚类和甲醛类化合物等能导致各种健康和环境问题的产品。

绿色遮阳板——万科中心采用耐桑公司生产的绿色遮阳纤维——一种无聚氯乙烯、无挥发性产品。产品不仅在使用期间不会挥发有机气体，还能够更快的回收和转移到填埋坑。

可持续场地

该建筑位于一片经改造过的稳定场地上，是市政雨水处理系统的一部分。潟湖起到了生态沼泽和蓄水池的功能，连接了附近的一些溪流。由斯蒂文·霍尔建筑事务所设计的景观水岸方案对市政硬景观进行了重新设计，将其打造成了软缘植被河口。作为一个恢复性的生态设施，万科中心景观设计与传统开发模式共同起到了保护本地生态系统、最小化地面径流、侵蚀和环境破坏的作用。

项目既是建筑又是景观，巧妙地将工程设施与自然环境交织在一起。通过将建筑抬离地面，一个开放的公园在相对封闭和私密的社区中打造了全新的社交空间。

场地总面积约60,000平方米；其中45,000平方米为绿化空间。加上主建筑的绿化屋顶（约15000平方米）——项目的总绿化面积与开发前的场地总面积基本持平。

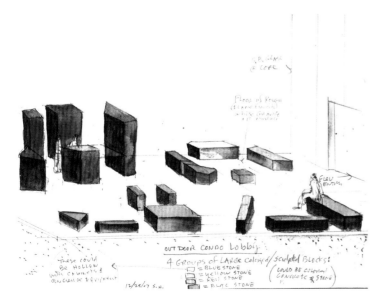

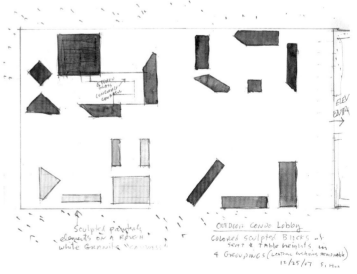

Sections:
1. Grass roof
2. Exposed concrete ceiling
3. Operable window
4. Exterior shading
5. Aluminium panels
6. Raised floor
7. Exposed concrete radiant ceiling

剖面图：
1. 植草屋面
2. 暴露混凝土天花
3. 可开启窗户
4. 室外遮阳
5. 铝板
6. 架高地板
7. 暴露混凝土辐射楼板

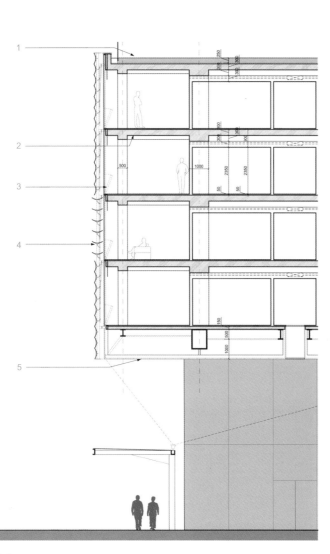
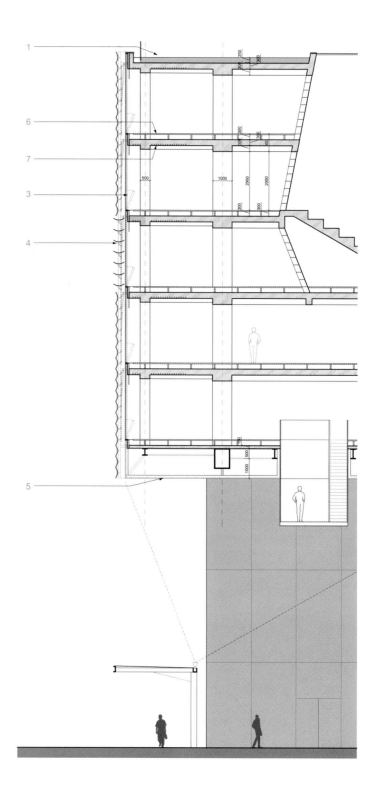

5. Elevation
5. 外立面

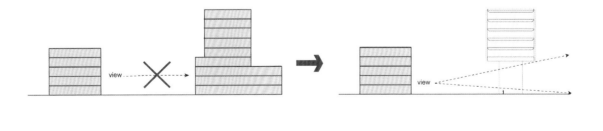

Maximised Views and Landscape
最大化视野与景观

203

**Wall Section at Parapet
(This Page and Facing Right):**
1. Operable perforated anodised aluminium louvres 275mm×2000mm
2. Anodised aluminium louvre bracket
3. Painted steel louvre frame
4. Painted steel bracket
5. Sensor-controlled hydraulic piston
6. Operable window
7. Insulating glazing unit with solar control coating
8. Curtain wall frame
9. Aluminium cover with fire insulation infill
10. Painted steel catwalk
11. Insulating glazing unit with acid-etched fire resistant glass
12. Room darkening shade
13. Floor diffuser
14. Modular raised flooring system
15. Façade automation control box monitors interior/exterior environment sensor
16. Aluminium coping
17. Waterproofing membrane
18. Drain
19. Modular planting system
20. Insulating glazing unit
21. Metal panel shadow box
22. Painted aluminium soffit panels

墙体剖面图
（本页图和对页右图）：
1. 可控式穿孔电镀铝百叶 275毫米×2000毫米
2. 电镀百叶支架
3. 喷漆钢百叶框架
4. 喷漆钢支架
5. 传感器控制压塞
6. 可控式窗户
7. 配有日光控制涂层的保温玻璃
8. 幕墙框架
9. 铝包层内部是防火隔热填充物
10. 喷漆钢天桥
11. 酸腐蚀防火保温玻璃元件
12. 室内深化遮阳
13. 地板分散器
14. 模块化高架地板系统
15. 外立面自动控制箱监视着室内外环境传感器
16. 铝顶
17. 防水膜
18. 排水
19. 模块化种植系统
20. 保温玻璃元件
21. 金属板阴影盒
22. 喷漆铝底板

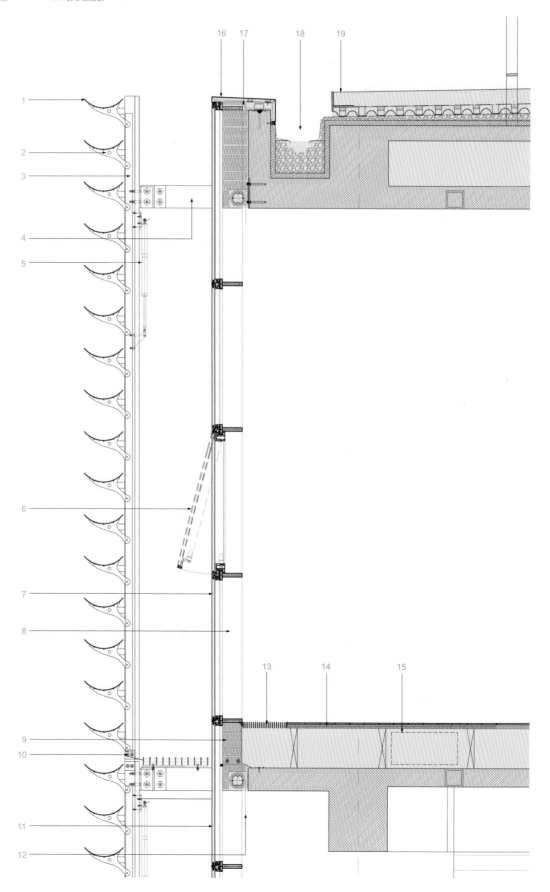

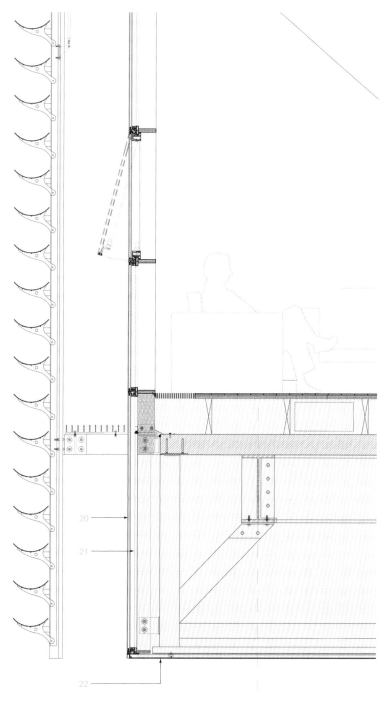

6, 7. Glass shadow
6、7. 玻璃阴影

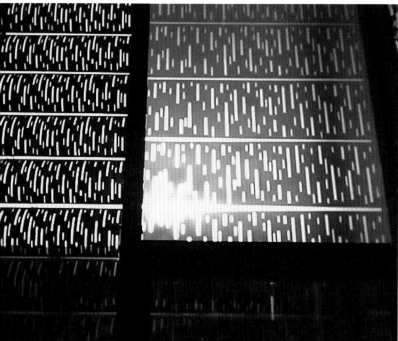
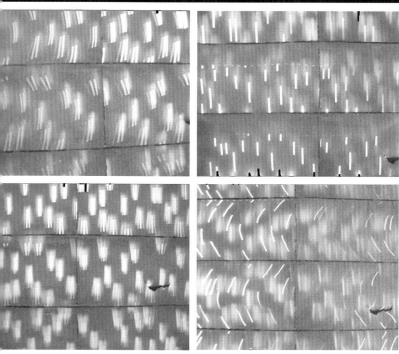
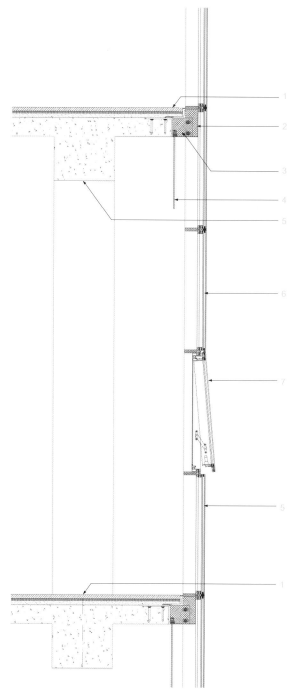

Curtain Wall (Above):
1. Floor finish
2. Aluminium panel shadow box
3. Fire insulation infill
4. Fire separation laminated glass
5. Exposed concrete
6. Exterior glass with steel back-up behind
7. Operable window

帘墙（上图）：
1. 楼板完成面
2. 铝制遮阳帘收纳盒
3. 防火绝缘材填充
4. 防火分区层压玻璃
5. 曝露混凝土
6. 背钢支撑之室外结构玻璃
7. 开启式窗户

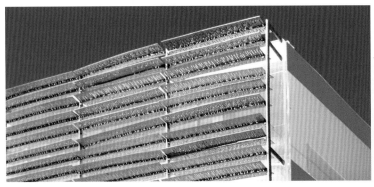
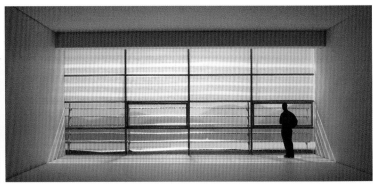

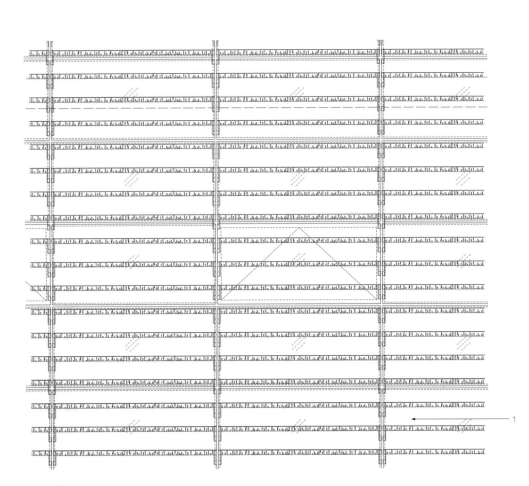

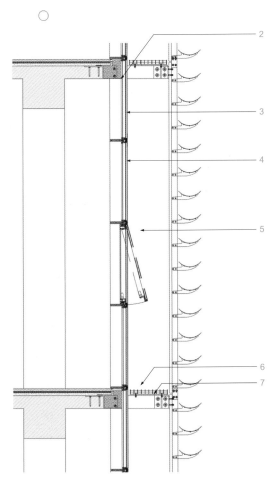

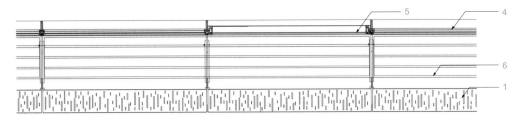

**Fixed Louvres
(These 3 and facing Below 3):**
1. Exterior shade panel
2. Aluminium panel shadow box
3. Fire separation laminated glass
4. Exterior structural glazing with steel back-up behind
5. Operable window
6. Steel grating catwalk
7. Steel bracket

固定百叶窗
（本页3图和对页下3图）：
1. 可开启式室外遮阳板
2. 铝制遮阳帘收纳盒
3. 防火分区层压玻璃
4. 背钢支撑之室外结构玻璃
5. 开启式窗户
6. 钢格栅过道
7. 钢制托座

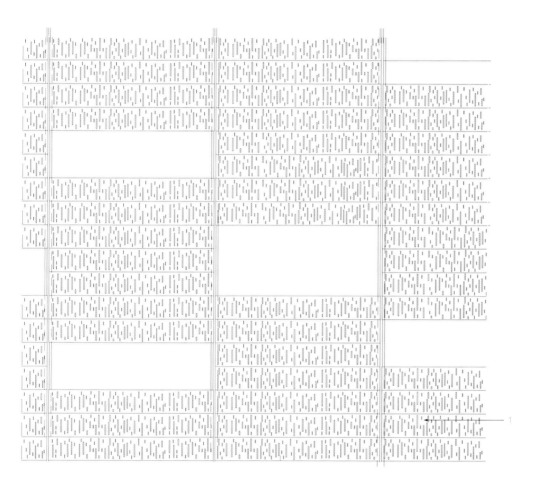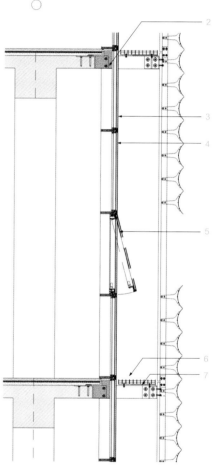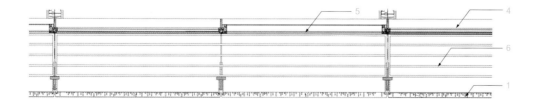

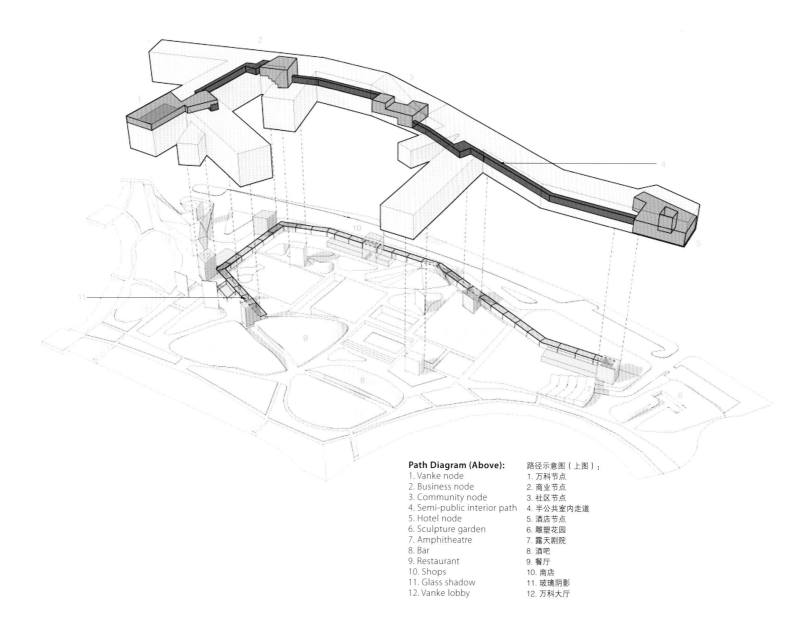

Path Diagram (Above):
1. Vanke node
2. Business node
3. Community node
4. Semi-public interior path
5. Hotel node
6. Sculpture garden
7. Amphitheatre
8. Bar
9. Restaurant
10. Shops
11. Glass shadow
12. Vanke lobby

路径示意图（上图）：
1. 万科节点
2. 商业节点
3. 社区节点
4. 半公共室内走道
5. 酒店节点
6. 雕塑花园
7. 露天剧院
8. 酒吧
9. 餐厅
10. 商店
11. 玻璃阴影
12. 万科大厅

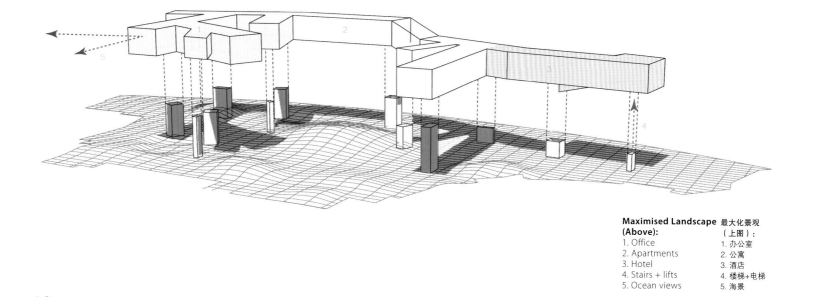

Maximised Landscape (Above):
1. Office
2. Apartments
3. Hotel
4. Stairs + lifts
5. Ocean views

最大化景观（上图）：
1. 办公室
2. 公寓
3. 酒店
4. 楼梯+电梯
5. 海景

8. Day view
9. Night view

8. 白天景观
9. 夜景

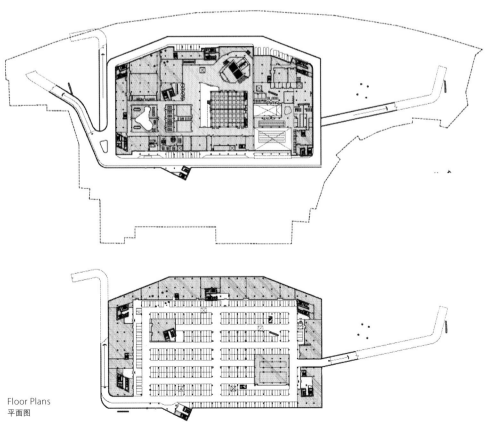

Floor Plans
平面图

10. Interior
11. Stairs
12. Lifts
10. 室内
11. 楼梯
12. 电梯

LINKED HYBRID
Beijing
Steven Holl Architects

北京当代万国城

北京
斯蒂文·霍尔建筑事务所

Site Area: 220,000m²
Completion Time: 2008
Architect: Steven Holl Architects
Photographer: Andy Ryan, Iwan Baan, Shu He
Award: 2009 "Best Tall Building" in Asia & Australia Category

占地面积：220000平方米
建成时间：2008年
建筑设计：斯蒂文·霍尔建筑事务所
摄影师：安迪·瑞安，伊万·班，舒赫
奖项：2009年美国世界高层建筑协会亚洲与澳大利亚年度最佳高层建筑奖

The 220,000-square-metre Linked Hybrid complex includes eight towers linked by a ring of eight skybridges housing a variety of public functions. The complex is located adjacent to the former city perimeter of Beijing. To counter current urban development trends in China, the complex forms a new 21st century porous urban space, inviting and open to the public from every side. In addition to more than 750 apartments, the complex includes public, commercial, and recreational facilities as well as a hotel and school. With sitelines around, over, and through multifaceted spatial layers, this "city within a city" has – as one of its central aims – the concept of public space within an urban environment, and can support all the activities and programmes for the daily lives of over 2,500 inhabitants.

The ground level offers a number of open passages for all people (residents and visitors) to walk through. These passages include "micro-urbanisms" of small scale shops which also activate the urban space surrounding the large central reflecting pond. On the intermediate level of the lower buildings, public roof gardens offer tranquil green spaces, and at the top of the eight residential towers private roof gardens are connected to the penthouses. All public functions on the ground level, including a restaurant, hotel, Montessori school, kindergarten, and cinema, have connections with the green spaces surrounding and penetrating the project. Lifts displace like a "jump cut" to another series of passages on higher levels. From the 18th floor a multi-functional series of skybridges with a swimming pool, a fitness room, a café, a gallery, etc. connects the eight residential towers and the

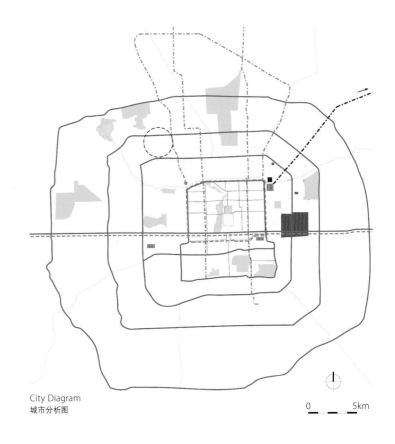

City Diagram
城市分析图

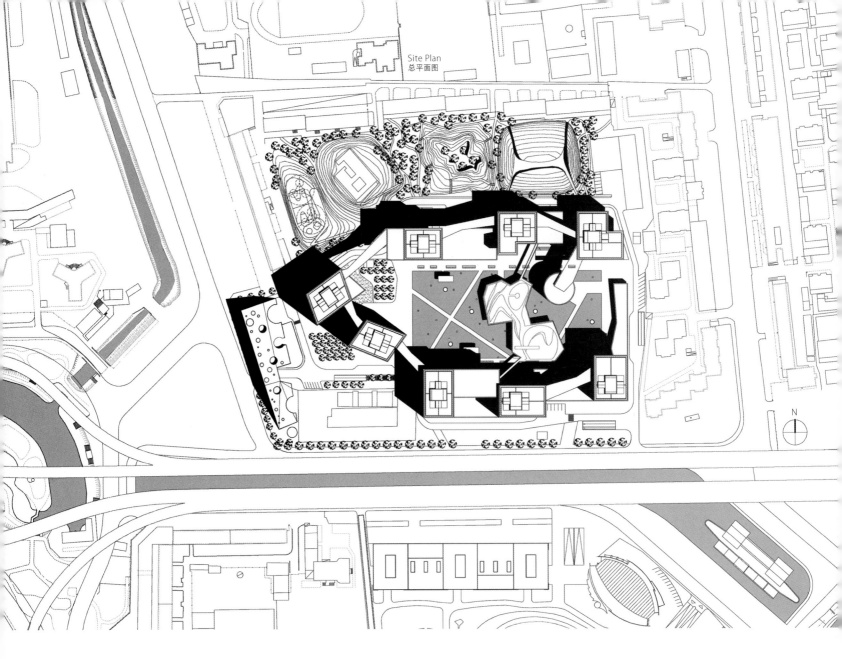

Site Plan
总平面图

hotel tower, and offers views over the unfolding city. Programmatically this loop aspires to be semi-lattice-like rather than simplistically linear. The designers hope the public sky-loop and the base-loop will constantly generate random relationships; functioning as social condensers in a special experience of city life to both residents and visitors.

Focused on the experience of passage of the body through space, the towers are organised to take movement, timing and sequence into consideration. The point of view changes with a slight ramp up, a slow right turn. The encircled towers express a collective aspiration; rather than towers as isolated objects or private islands in an increasingly privatised city, the designers' hope is for new "Z" dimension urban sectors that aspire to individuation in urban living while shaping public space.

Geo-thermal wells (660 at 100 metres deep) provide Linked Hybrid with cooling in summer and heating in winter, and make Linked Hybrid one of the largest green residential projects. The large urban space in the centre of the project is activated by a greywater recycling pond with water lilies and grasses in which the cinematheque and the hotel appear to float. In the winter the pool freezes to become an ice-skating rink. The cinematheque is not only a gathering venue but also a visual focus to the area. The cinematheque architecture floats on its reflection in the shallow pond, and projections on its façades indicate films playing within. The ground floor of the building, with views over the landscape, is left open to the community. The polychrome of Chinese Buddhist architecture inspires a chromatic dimension. The undersides of the bridges and cantilevered portions are coloured membranes that glow with projected nightlight and the window jambs have been coloured by chance operations based on the "Book of Changes" with colours found in ancient temples.

The water in the whole project is recycled. This greywater is piped into tanks with ultraviolet filters, and then put back into the large reflecting pond and used to water the landscapes. Re-using the earth excavated from the new construction, five landscaped mounds to the north contain recreational functions. The "Mound of Childhood", integrated with the kindergarten, has an entrance portal through it. The "Mound of Adolescence" holds a basketball court, a roller blade and skate board area. In the "Mound of Middle Age" we find a coffee and tea house (open to all), a Tai Chi platform, and two tennis courts. The "Mound of Old Age" is occupied with a wine tasting bar and the "Mound of Infinity" is carved into a meditation space with circular openings referring to infinite galaxies.

1. Façade
2. Full view of façade
1. 外立面
2. 外立面全景

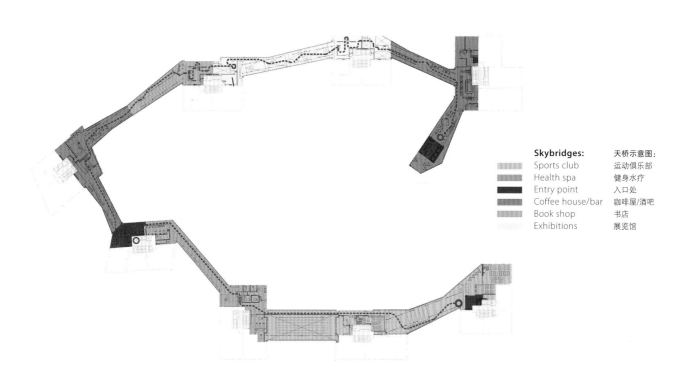

	Skybridges:	天桥示意图:
	Sports club	运动俱乐部
	Health spa	健身水疗
	Entry point	入口处
	Coffee house/bar	咖啡屋/酒吧
	Book shop	书店
	Exhibitions	展览馆

3. Tower and cinematheque/tea house
4. Entrance of tea house
3. 塔楼和实验电影院/茶室
4. 茶室入口

北京当代万国城占地220,000平方米，包括8栋环环相连的建筑，通过人行天桥连成一个环形，天桥有多种功能。这个建筑群坐落在北京市原城市边缘区。为了跟中国当前的城市化风格形成对比，该建筑群打造出一种新型的21世纪具有穿透力的城市空间，从外面各个方向来看都是开放的且充满诱惑力。建筑群除了包含750个住宅单元，还包括公共空间、商业区、休闲娱乐设施等，此外还有一家酒店和一所学校。这座"城中城"空间层次分明，功能多种多样，其宗旨是要成为城市环境中的一个概念上的公共空间，在这里能够进行各种活动，为这里的2500名居民的日常生活提供良好的服务。

一楼为所有人（包括住户和参观者）提供了许多开放的走道。这些走道可以说是"微型城市规划"，包括小型商店，这些商店为中央碧波荡漾的池塘周围的城市空间注入了活力。比较低矮的建筑里，在中层高度上都有屋顶公共花园，这里静谧的绿色空间让人心旷神怡。8座住宅高塔的顶层则是私人屋顶花园，跟大楼平顶上的阁楼相连。一楼所有的公共功能，包括餐厅、酒店、蒙台梭利学校、幼儿园、电影院，都跟贯穿整个项目的周围绿色空间相连。乘坐电梯，仿佛电视片的"跳跃剪辑"，就来到高层的另一系列的走道。从18层往上，一系列多功能的人行天桥、一个游泳池、一间健身房、一家咖啡馆、一个画廊等等，把8座住宅高塔跟酒店大楼连接起来，而且这些地方都能为我们提供望向这座伸展着的万国城的不同视角。这种环形结构是一种半框架设计，而不是简单的线性结构。设计师希望公共"天环"和"地环"能够相互关联，而这种关联又是随意的；此外这里还试图让居住者和参观者都能体验一种独特的城市生活体验，成为一种"社会冷凝器"。

各个高塔的设计尤其关注建筑实体之间的空间过渡，让这些建筑能够有节奏、有序地动起来。随着斜坡的缓缓上升和一个右转，视线会跟着发生变化。围成环形的这些建筑表达了一种集体意识。设计师特意避免将建筑设计成相互分离的单独建筑个体，因为现代社会城市已经越来越个体化，设计师希望借助这样一座"城中城"的设计，能够既让人保持独立性，又兼顾公共空间。

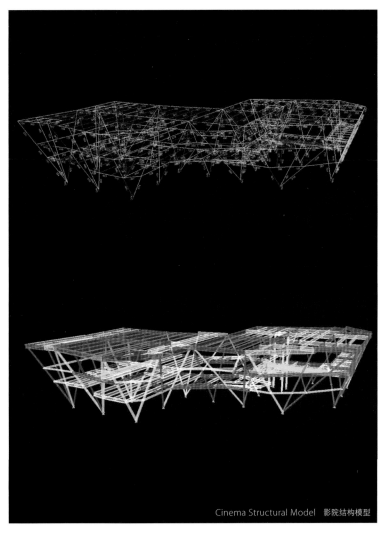

Cinema Structural Model 影院结构模型

4

地热井（660个，在地下100米深处）为北京当代万国城提供了冬暖夏凉的环境，也让万国城成为最大的绿色住宅项目之一。项目中间宽敞的公共空间中有一个灰水循环处理池塘，里面种着睡莲和水草，让这个空间充满活力，电影院和酒店仿佛都漂浮起来了。冬天池塘里的水结冰以后，这里就变成一个滑冰场。电影院不但是人们聚会的场所，而且是这里的一个视觉焦点。这座电影院建筑的影像倒映在池塘浅浅的水面上，随着微波荡漾，建筑外立面上会通过投影显示里面正在放映的影片。这座建筑的一楼可以看到外面的风景，对这个社区居民开放。中国佛教彩饰的运用为这里的色彩添加了亮丽的一笔。天桥和悬垂部分的底部都镀上一层色彩，夜晚会随着投射的灯光熠熠生辉。窗户侧柱根据《易经》随机涂上各种颜色，这些颜色都是在古庙中常见的颜色。

整个项目用水都是循环利用的。灰水通过管道进入紫外线过滤水槽中，然后被放回到倒影池，用于水景景观。挖掘出的土被重新应用于新工程，在北面打造了五座景观高地。"童年高地"与幼儿园相结合；"青少年高地"设有篮球场和轮滑与滑板区；"中年高地"有咖啡馆兼茶室（面向所有人开放）、太极平台和两个网球场；"老年高地"上是一个品酒吧；"无限高地"则被打造成一个冥想空间，圆形开口象征无限的银河。

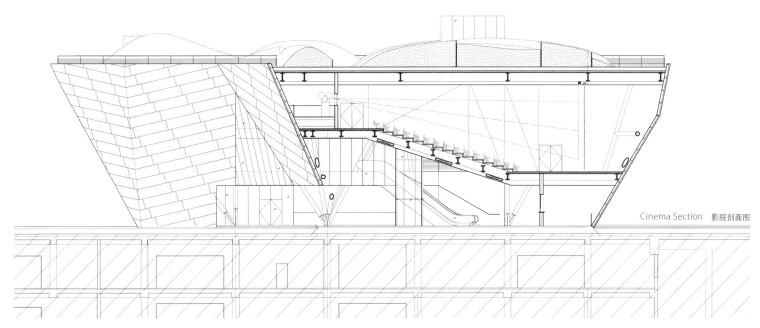

Cinema Section 影院剖面图

5. Group exercise space
5. 运动空间

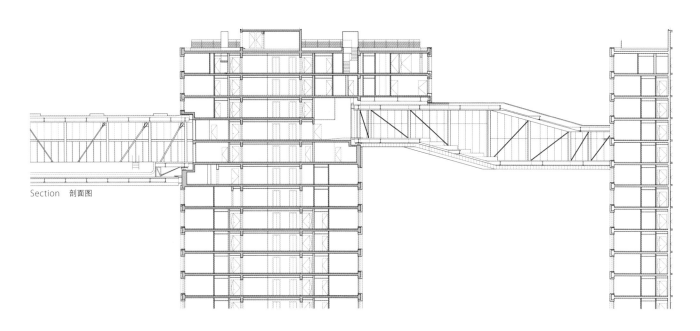

Section 剖面图

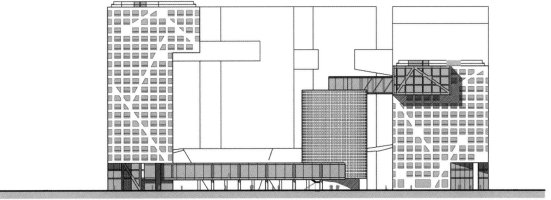
Elevation 立面图

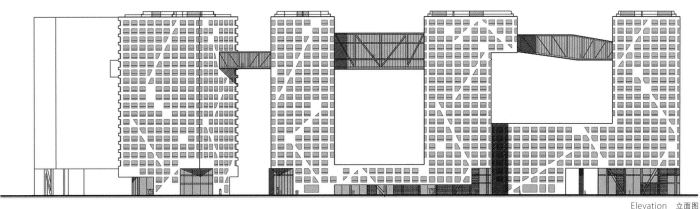
Elevation 立面图

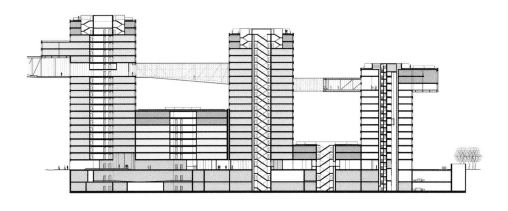

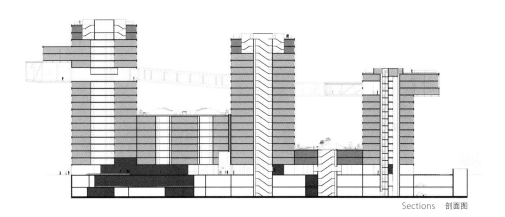
Sections 剖面图

221

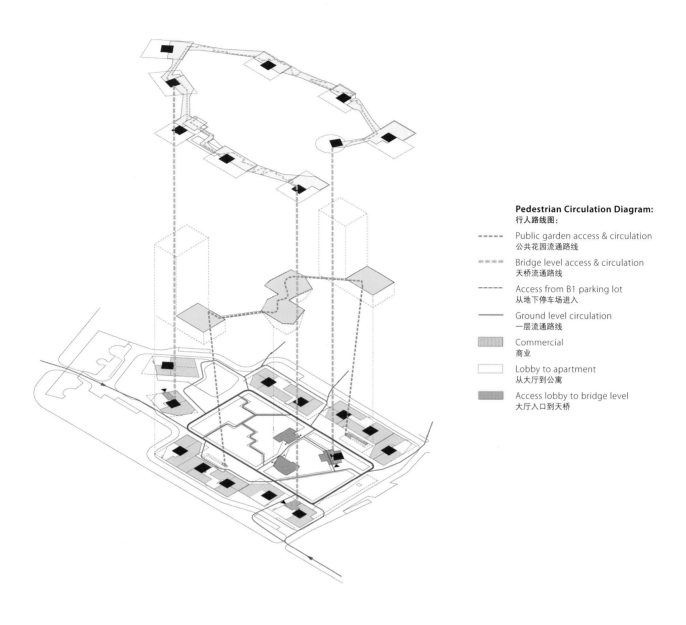

Pedestrian Circulation Diagram:
行人路线图：

- - - - Public garden access & circulation
公共花园流通路线

- - - - Bridge level access & circulation
天桥流通路线

- - - - Access from B1 parking lot
从地下停车场进入

——— Ground level circulation
一层流通路线

▨ Commercial
商业

☐ Lobby to apartment
从大厅到公寓

▨ Access lobby to bridge level
大厅入口到天桥

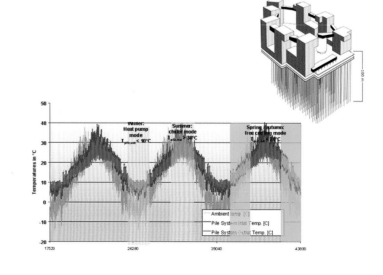

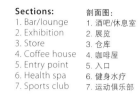

Sections:
1. Bar/lounge
2. Exhibition
3. Store
4. Coffee house
5. Entry point
6. Health spa
7. Sports club

剖面图：
1. 酒吧/休息室
2. 展览
3. 仓库
4. 咖啡屋
5. 入口
6. 健身水疗
7. 运动俱乐部

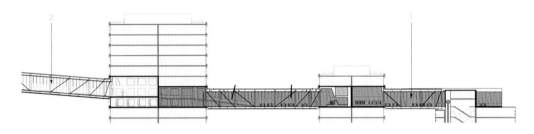

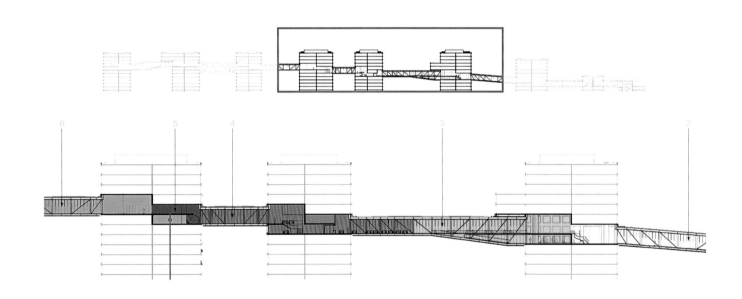

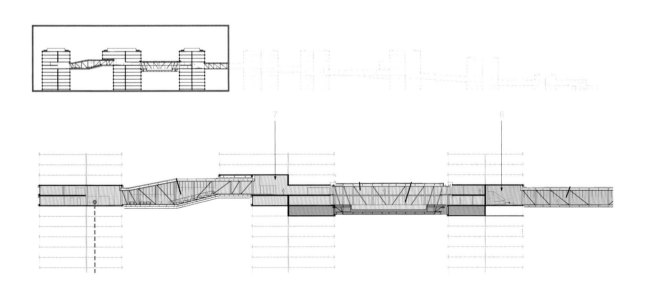

6. Corridor
7. Interior
8. Gallery
6. 过道
7. 室内
8. 画廊

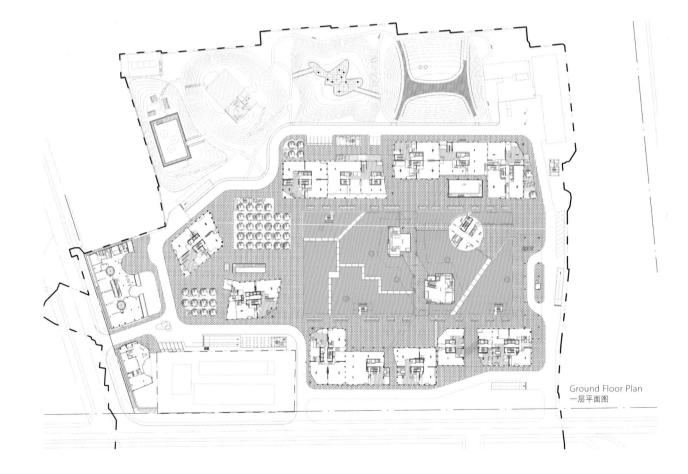

Ground Floor Plan
一层平面图

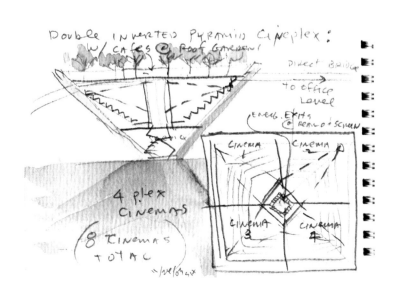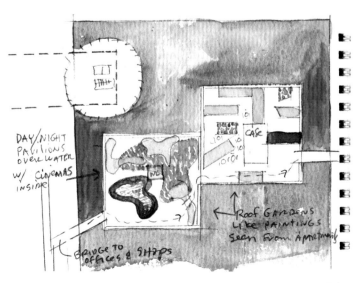

CHINA NATIONAL OFFSHORE OIL CORPORATION
Beijing
Kohn Pedersen Fox Associates PC

中国海洋石油总公司
北京
KPF建筑师事务所

Gross Floor Area: 90,000m²
Completion Time: 2006
Architect: Kohn Pedersen Fox Associates PC
Photographer: Kohn Pedersen Fox Associates, ZHANG Guangyuan, H.G. Esch
Client: CNOOC
Awards: 2007 AIA Hong Kong - Architecture Merit Award and Sustainability in Design Award, 2003 Design Excellence Award, Beijing Urban Planning Commission

建筑面积：90000平方米
建成时间：2006年
建筑设计：KPF建筑师事务所
摄影师：KPF建筑师事务所，张光远，H·G·埃施
业主：中国海洋石油总公司
奖项：美国建筑师协会香港分会2007环保建筑设计优异奖；
北京城市规划委员会2003优秀设计奖

Located at a major crossroads along the Second Ring Road in Beijing's Dongcheng District, the CNOOC headquarters building acts as an urban counter-point to the massive Ministry of Foreign Affairs Building situated on the opposing corner. The building's form evokes the images of offshore oil production. The prow-like shape recalls an oil tanker's bow, and the tower mass elevated above the ground on piloti suggests an offshore oil derrick. This effect is further heightened by the design of the the ground plane which has been developed to suggest the ocean's surface. The rotated triangular tower maximises the use of the site and creates an entry courtyard along the quieter side which is entered through a symbolic gateway recalling traditional Chinese courtyards.

Internally, office and function spaces are organised around a central, full-height, sunlit atrium. Large sky-gardens carve away portions of the tower floorplates to allow daylight to penetrate into the atrium from all three sides. These sky-gardens take on different configurations on each of the sides in response to the sun angles encountered. Additionally, a skylight and clerestory windows at the top of the atrium allow filtered light to wash the atrium interior surfaces.

In summary, the proposed design will provide for CNOOC a unique iconic structure that symbolises their corporate mission and will give to the city of Beijing a strong and memorable landmark for its citizens.

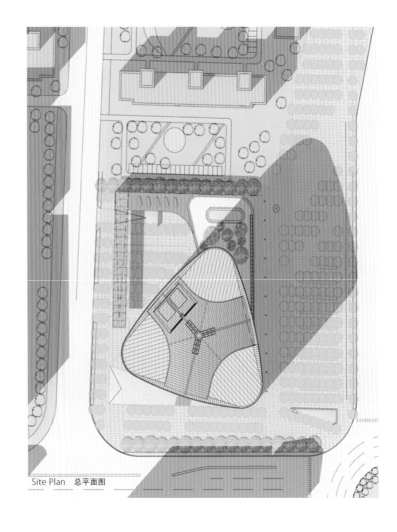

Site Plan 总平面图

中国海洋石油总公司坐落于北京东城区二环路沿线的十字路口处,与对面的外交部大楼相互呼应。该建筑形式让人联想到对海上石油生产。裙房船头形状好似海上钻井平台,而高层的形状又好似高高的钻机,犹如海平面一般的地面设计更突出了这一效果。带有一定弧度的三角形造型充分利用了地块面积,并沿幽静的一侧打造了入口庭院,让人不禁想到传统的中式院落。

内部办公区与其他功能区沿着与建筑等高且光线充裕的中央心房展开,空中花园似乎从不同楼层之间雕塑出来一般,光线由此照射到心房内的各个角落。空中花园从不同的角度望去呈现不同的造型,并随光线的角度而变化。此外,心房上方的天窗以及通风窗进一步将光线引入进来。

简言之,这一设计赋予中国石油总公司一个独特的形象,既能体现出其功能性,又为北京市民打造了一个永恒的地标建筑。

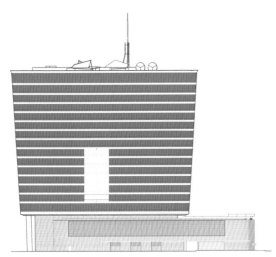

East Elevation 东立面图

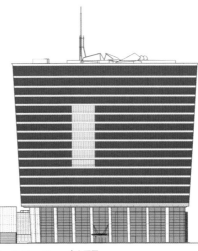

South Elevation 南立面图

West Elevation 西立面图

1. Upward view
2. West elevation
3. Garden
1. 仰拍图
2. 西立面
3. 花园

Section:
1. Office
2. Atrium
3. Sky-garden
4. Auditorium
5. Meeting room
6. Service
7. Lobby
8. Health Club
9. Parking

剖面图：
1. 办公室
2. 中庭
3. 空中花园
4. 礼堂
5. 会议室
6. 服务区
7. 大堂
8. 健身俱乐部
9. 停车场

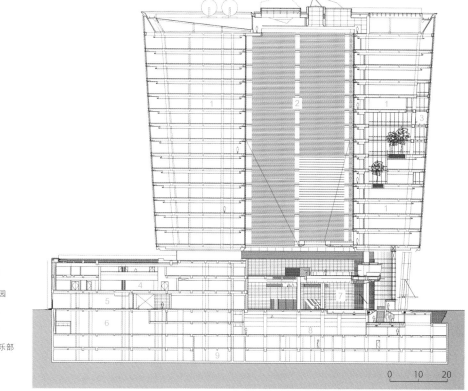

4. East elevation
5-7. Façade details
8, 9. Lobby
4. 东立面
5-7. 墙皮
8、9. 大厅

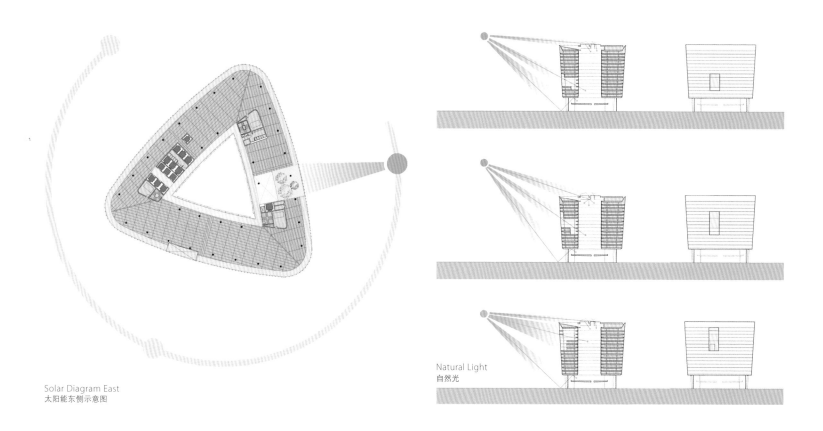

Solar Diagram East
太阳能东侧示意图

Natural Light
自然光

10

11

10, 11. Atrium
10、11. 中庭

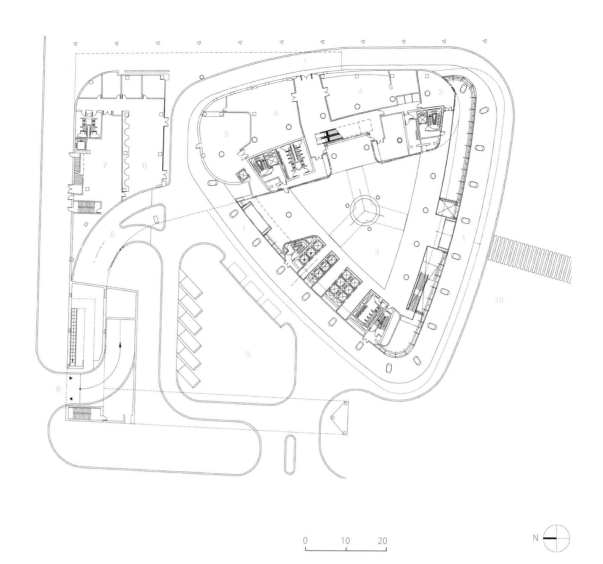

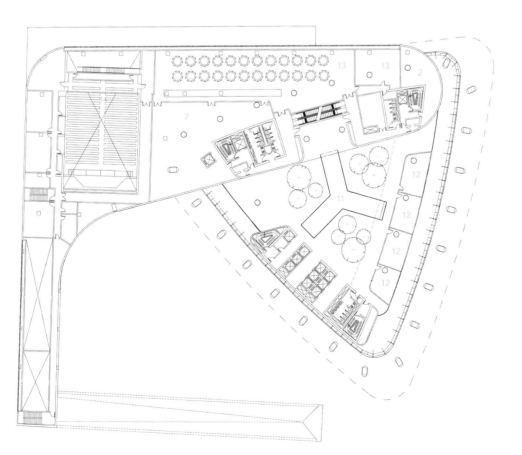

Ground Floor Plan (Above) and Third Floor Plan (Left):
1. Entry
2. Mechanical
3. Lobby
4. Exhibition
5. Shop
6. Café
7. Kitchen
8. Garage entry
9. Drive court
10. Garden
11. Atrium
12. Meeting rooms
13. Corporate
14. Auditorium

一层平面图（上图）
和四层平面图（左图）：
1. 入口
2. 机房
3. 大堂
4. 展览
5. 商店
6. 咖啡厅
7. 厨房
8. 车库入口
9. 停车场
10. 花园
11. 中庭
12. 会议室
13. 公司办公室
14. 礼堂

BEA FINANCIAL TOWER
Shanghai
TFP Farrells

东亚银行金融大厦
上海
TFP建筑设计事务所

Gross Floor Area: 70,000m²
Design/Completion Time: 2004/2009
Architect: TFP Farrells
Photographer: Paul Dingman Photo, ZHOU Ruogu
建筑面积：70000平方米
设计/建成时间：2004年/2009年
建筑设计：TFP建筑设计事务所
摄影师：保罗·丁曼摄影公司，周若谷

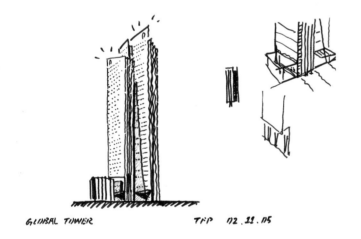

TFP's designs for the BEA Financial Tower combine elegant, contemporary aesthetics with a modern, technologically advanced building form. The striking development offers high efficiency levels and responds to China's increasing concern for environmental protection.

Although there are view corridors of the river and the Bund, the site is set back from the waterfront and has to compete with prominent high rises, notably the Jin Mao Building, currently the tallest skyscraper in China. TFP proposed a structure that was layered into three principal forms. A central circulation and service core is flanked by two floor plates with the west wing of the building rising above the other two components. The creation of this stepped effect brings a level of clarity and directness to the building's massing.

Each element functions independently but is bound into a singular composition by complementary materials and modularity. This adds significantly to BEA Financial Tower's instant-recognition factor and enhances both the view potential and the building's silhouette on the skyline.

The façades react differently to the environment through orientation, materials and technology within the building envelope. Fluctuations in heat gain and loss are limited, the building's sustainability is maximised, and operational efficiency is improved.

Following detailed analysis into solar insulation, four types of cladding were established, each of which is designed to deal with a specific environmental aspect. To minimise excessive solar gain and building heat load on the southwest and southeast elevations, the percentage of glazed areas is reduced, horizontal shading devices are provided and low-emissive glass is used. Glare protection from the low-setting sun is required on the northwest side but because the principal views of the local park lie in this direction maintaining an open, glazed vista is important. To keep the views while simultaneously limiting glare, large areas of the northwest façade are glazed and vertical fins introduced to shade the interior and allow for lighting effects at night. As well as producing an elongating effect, the fins add visual interest and depth to the façade rendering at all times.

The northeast façade has neutral solar gain and is mainly glazed with low-e glass to maximise views. TFP captured the sense of the greenery being swept vertically into the building by positioning sky gardens on the various refuge floors and creating a visual link to the park from ground level upwards.

1. Façade detail
2. Façade
1. 外墙细节
2. 全景

The use of other energy-efficient systems was also a main intent of the design and the building has a responsive Building Management System (BMS). This controls the interior environment to achieve optimum use of energy resources and maintains the internal air temperature and air quality. Solar collectors are set on an angled surface on the roof of the southwest block of the tower to achieve optimum performance, contribute to the lighting of the common areas and potentially pre-heat the air and water systems. Grey-water collection from the roof is also utilised for irrigation and flushing-water purposes.

During the building's lifetime, the net aggregate of all these systems will contribute to the limitation of energy use and enhance the profile of the development as an environmentally aware and responsible contribution to the skyline of Pudong.

这一建筑集典雅、现代美学及融合先进技术的造型于一身，达到高水准要求的同时，更重要的是它满足了中国当前时代日益关注的有关绿色理念要求。

建筑选址在远离河岸处，与中国目前最高的建筑，金茂大厦相"抗衡"。三种造型层叠在一起，中央通道及服务中心通过楼板"挂"在西侧结构上，似乎从其他两个结构上"冉冉升起"一般。阶梯状造型突出了层次感及简约性。

相互补充的材质以及模块结构的运用使得每一种元素在行使自身功能的同时，又依附于整体。这一设计增添了建筑本身的可识别性并增强了其潜在的"能量"及形象。

外观通过朝向、材质以及内部技术的运用可应对不同的环境变化，热量的获取及损失的变化被缩小，可持续性被放大，从而实现了能源节约。通过对太阳绝缘材料的仔细研究，四种覆层结构被运用，分别应对不同的环境条件。为减少西南及东南两侧太阳热量的获取，玻璃覆层的运用被大量减少，遮光结构以及低反射玻璃被采用。

大厦的西北侧需要设置夕照遮阳设备，但是由于地方公园也设在这个方向，西北侧还要求保留开放通透视野。为了在保证视野的同时减少刺眼的阳光，西北立面的大部分都采用了玻璃装配，而垂直扇片则被用作为室内遮阳并在夜晚打造灯光效果。于此同时，扇片还产生了延伸效果，外立面添加了视觉冲击力和深度。

东北立面获得较多日照，大部分空间都采用低辐射玻璃来最大化清晰的视野。TFP通过在各个隔火层设置空中花园并与公园建立了视觉连接，为整座大厦带来了绿色效应。其他节能系统的使用也是设计的主要目的之一，建筑拥有一个应激性建筑管理系统。该系统控制着室内环境，以取得最佳的能源利用效果、保持室内温度和空气质量。

太阳能收集器被呈角度地设置在大厦西南侧的屋顶，以获得最佳绩效，为公共区域的照明和空气与水的预热系统提供了能量。屋顶收集的灰水则被运用到灌溉和洗手间冲水。

在建筑的使用期间，所有系统的净额总计都将对能源限制做出贡献，并且提升项目开发的环保价值，为浦东新区的建设负责。

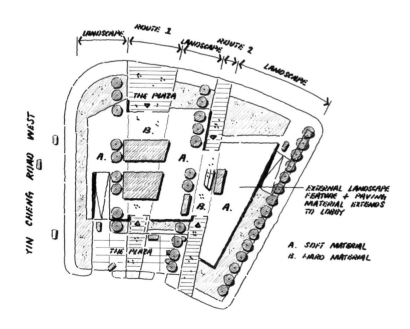

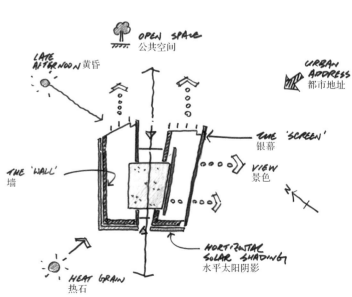

3. Night view of façade
4. Southeast elevation
5. Entrance
3. 夜景
4. 东南立面
5. 入口

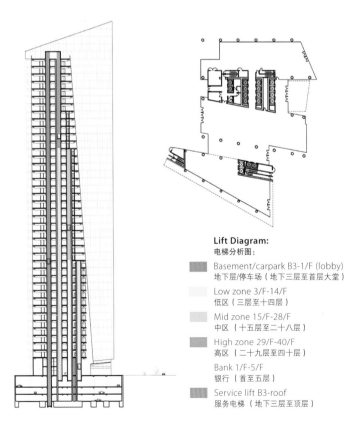
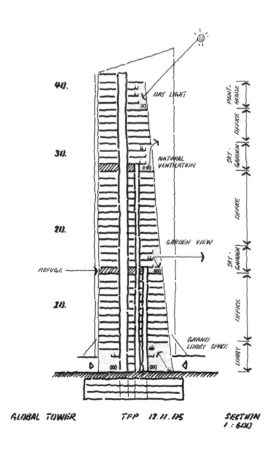

Lift Diagram:
电梯分析图：

- Basement/carpark B3-1/F (lobby)
  地下层/停车场（地下三层至首层大堂）
- Low zone 3/F-14/F
  低区（三层至十四层）
- Mid zone 15/F-28/F
  中区（十五层至二十八层）
- High zone 29/F-40/F
  高区（二十九层至四十层）
- Bank 1/F-5/F
  银行（首至五层）
- Service lift B3-roof
  服务电梯（地下三层至顶层）

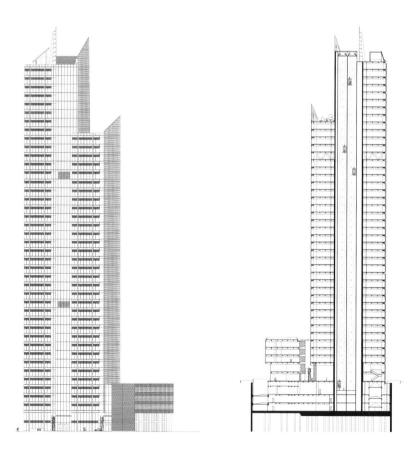
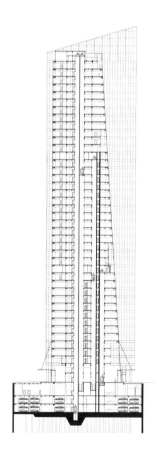

Sections
剖面图

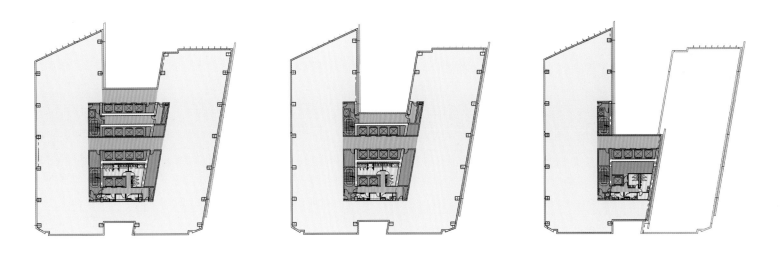

Office
办公空间

Lobby
大堂

Bathrooms
洗手间

Lift core & back-of-house service
电梯及后室

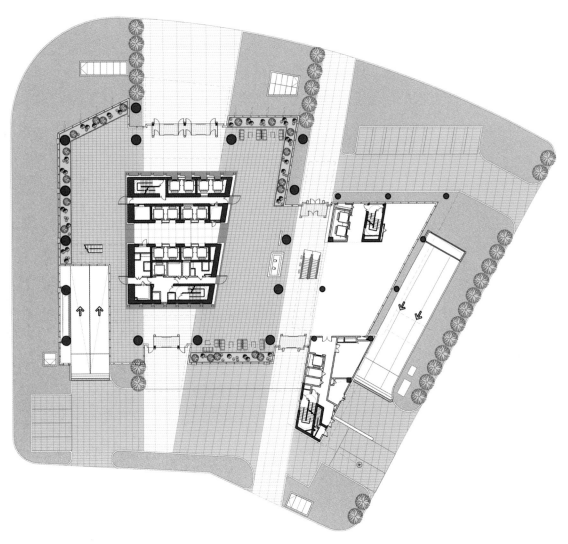

6, 7. Lobby
6、7. 大厅

RIVIERA TWINSTAR SQUARE
Shanghai
Arquitectonica

浦江双辉大厦

上海
Arquitectonica 建筑事务所

Site Area: 200,000m²
Design/Completion Time: 2011
Architect: Arquitectonica
Photographer: Rogan Coles

占地面积：200000平方米
设计/建成时间：2011年
建筑设计：Arquitectonica 建筑事务所
摄影师：罗根·高斯

Green Materials and Equipments Involved:
As a financial office building of the new era, Riviera TwinStar Square attaches great importance to environmental protection, energy conservation and sustainable development.

The façade comprises unitised glass curtain wall with natural stone and aluminium cladding. The façade materials were procured locally in China and are entirely recyclable. Attention was also paid to procured materials with low embodied energy, such as natural stone. The purchase of glass and the manufacturer of the stone material for the façade sourced close to the site, reduces pollution and energy consumption during the transportation process and maintains a thread towards sustainable design. Other materials were also locally sourced and procured.

The goal of lowering energy use during the service life of the building is achieved by using a high performance Low-E glass with solar control coating for the curtain wall.

The façade uses triple-pane Low-E glass: the outer pane is laminated glass, the middle is the air layer, and the inner layer is single-pane glass. The triple-pane insulated glass not only improves winter insulation performance, reduces summer heat gain, enhances acoustic performance of the wall but also mitigates the safety concerns of using tempered glass on a high-rise building. This way, it can aim to reach higher and improved safety and energy savings. The special low reflective coating on the glass, stone and metal elements projecting from the design further minimise light pollution to the adjacent environment. The use of stone cladding offers another layer of thermal insulation. This not only helps solve architectural design issues but helps augment the thermal performance of the façade and achieve sgreater sustainability targets.

Other Eco Features:
The air-conditioning uses full air system with double-route individual control on every floor, the exterior wall air-conditioning can be adjusted according to the season, and the interior frame-core tube air-conditioning can be fully utilised, for heat and for cold, saving energy significantly.

This building uses a centralised tube design. For example, unilateral lighting with shorter depth on the exterior wall, and lighting design has combined with the architectural space. Furthermore, the exterior lighting can be adjusted according to the sunlight intensity, to make reasonable lighting energy saving.

Elements of the energy system design – for the overall office complex, as well as the residential and hospitality building components – have been fully integrated, using techniques such as stagger the rush time, greatly reducing energy consumption.

The whole project is low-carbon energy driven. The choice of materials, use of recycled materials, or reuse of materials, can fully embody the concept of sustainable development.

The Story Behind the Building:
Riviera TwinStar Square is the first part of the Phase II Development of Lujiazui Financial District. As the first phase of development of Lujiazui Financial District in Pudong, Shanghai, it explores innovation based on the former design of existing office building complexes in the development zone. The project focused specifically to design a raised basement level of 3m that catered to the height of the Huangpu River's flood protection systems, enabling the building along the river to take full advantage of the landscape on the river, while the raised part being the platform for

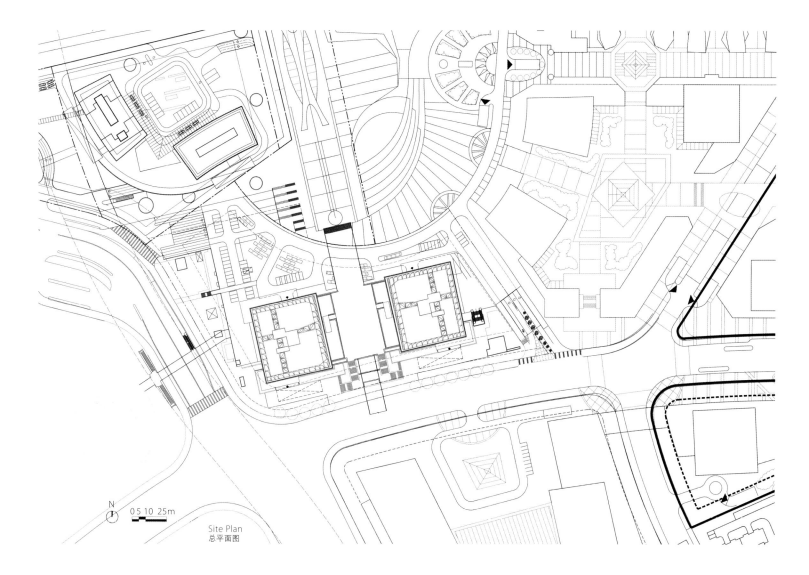

Site Plan
总平面图

people to walk and rest. The whole development zone is connected with the surrounding buildings by the platform, on which there is an outdoor pedestrian corridor, and an indoor one below the platform. A four-lane city road is built one level below the basement and is designed to connect the numerous sites within close proximity, thus greatly decreasing traffic congestion on the neighbouring surface roads.

The architectural design makes an overall integration aiming at the needs of financial office building. The location of the building becomes clearer by catering to the various requirements of the client. The typical office areas are smartly integrated with VIP office areas, taking into account a wide variety of different clients' needs. The structural design is well considered and accommodates the need of larger loads in areas such as mechanical and IT rooms. A removable slab area is reserved for unique tenant needs, allowing connectivity to upper and lower floors that adapt to the requirements of large and small financial institutions. Individual MEP areas are reserved to help ensure the high electricity supply requirements for financial institutions. The design includes considerations for individual or integrated electronic patrol, monitor, control, entrance guard and automatic reaction systems, which can be chosen freely by the clients according to their individual needs.

The conventional RC frame-core wall system is adopted for the tower to meet Client's programme and cost targets. However, some special structural features have been used for this project. Inclined columns are used at the facing curved elevations to integrate the structure into the architectural form. Steel Reinforced Columns (SRC) are introduced for the lower part of twin towers to minimise column sections for higher floor efficiency.

1. Overall view of façade
2. Building entry
3. Landscape
1. 全景
2. 入口
3. 景观

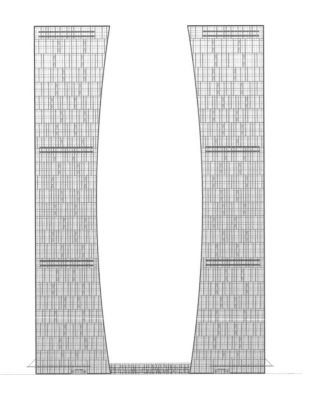
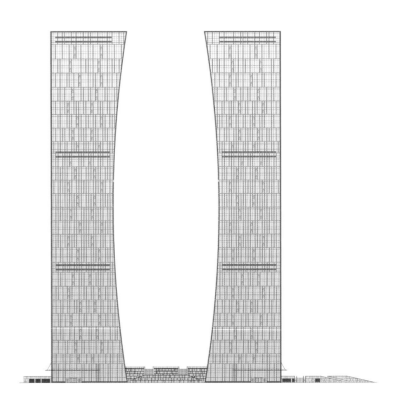

相关绿色材料与设备：

浦江双辉大厦作为新时期的金融办公大楼，对其环保、节能、可持续发展等性能给予了高度重视。

玻璃幕墙采用单元式系统，外侧安装天然石材和铝板装饰条。幕墙材料在中国本地生产，可完全回收利用，玻璃也在建筑所在的城市周边采购，减少了运输污染和能耗；建筑外墙装饰材料主要为石材，石材产地距离建筑较近，减少了运输过程中的污染，而天然石材在生产过程中的碳排放也较低；建筑内衬材料也使用当地的工业产品，有效地实现了对资源的再利用。玻璃幕墙采用镀有阳光控制膜的高性能Low-E玻璃，可以实现在建筑使用年限之内低能耗的设计目标。建筑外墙采用三层Low-E玻璃，外层采用夹胶玻璃、中间为空气层、内层为单片玻璃，不仅能提高幕墙系统的冬季保温和夏季隔热性能，获得更佳的隔声性能，还降低了在高层建筑使用钢化玻璃带来的风险，这样在安全性、节能上都满足了较高的要求。玻璃上镀有特殊的低反射膜，外部石材和金属装饰条有序排列，将玻璃幕墙对相邻建筑的光污染影响程度降至最低；石材饰面与外露的金属构件相比，具有更好的隔热性能，不仅有助于实现建筑师的设计理念，同时也提高了幕墙系统的保温性能，可进一步实现可持续发展的设计目标。

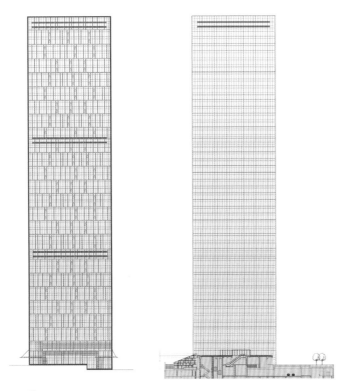
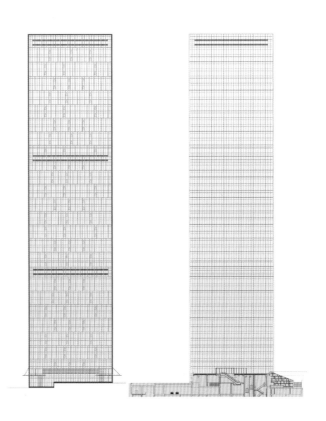

Elevations
立面图

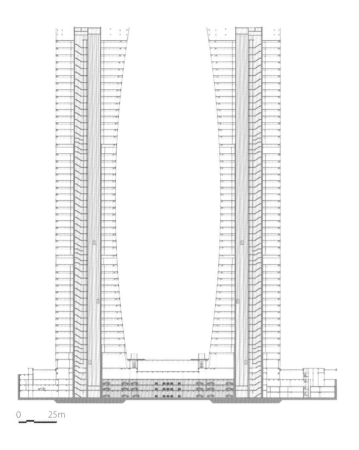

0 25m

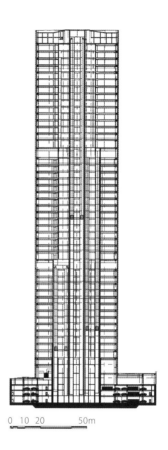

0 10 20   50m

4. Bird's-eye view
4. 鸟瞰图

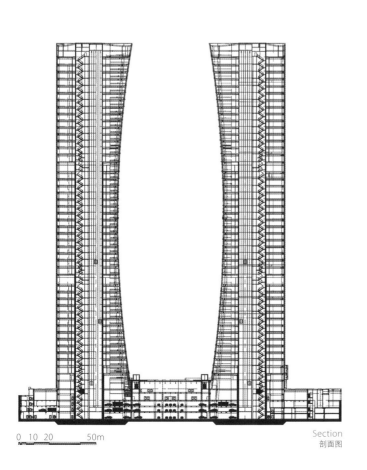

0 10 20   50m

Section
剖面图

249

5, 6. Skin detail
5,6. 外墙细节

其他生态特点：

空调系统采用全空气系统，每个楼层实行双回路独立控制，外墙空调可根据季节调节，内层核心筒空调可充分利用余热余冷，大量节约能源。

建筑采用了中心筒设计，外墙单侧采光，进深较少，同时灯光结合建筑空间：外侧灯光可以根据日光强度调节，从而合理地节约照明能源。

空调采用先进的净化系统，高于中国普通标准的换气量，同时每个风口、风道都设有完善的自净化装置，提高了空调的使用效率和寿命。外墙设有可开启扇，增强了通风效果，降低了空气污染的几率。

办公楼整体建筑和一期酒店公寓等其他建筑的部分能源系统采用整合设计，利用错时高峰等技术手段大大减少了能源消耗。

整个建筑低碳节能，材料选用或是采用二次循环材料或是可以二次循环利用，充分体现了可持续发展的理念。

建筑背后的故事：

浦江双辉大厦是陆家嘴金融中心区二期项目的第一期项目，做为上海浦东陆家嘴金融中心区的新一阶段的发展，其在原有开发区办公楼群设计的经验基础上进行了创新。

将整个区域地下一层提高3米，整体建筑基底和黄埔江防汛高度一致使沿江建筑充分利用了江面景观，同时抬高的建筑基底平面成为日常生活休息的平台。该区域也利用平台将周边建筑相互联系，在平台上形成室外人行连通走廊，在平台下形成了室内人行连通走廊。平台下方地下一层设有连通整个区域的四车道城市道路，大大减少了周边城市地面道路的压力。

建筑设计针对金融类办公楼的需求，进行了整体整合，在设计阶段考虑了不同客户的使用需求，使得建筑物定位更为清晰，减少客户入住后的改造。建筑功能分区上每个分区都设有专用的交易楼层、会议楼层，针对不同客户群也设置了普通办公区、VIP办公区。结构上考虑了金融机构可能的电子机房、档案房等大负载区域，预留了大中型金融机构上下层连通时需要的可拆除楼板区。机电上考虑了对于有较高供电保障的金融机构预留的独立发电机区域，考虑了独立或综合的电子巡更、监管、控制、门禁、自动反应等系统。客户入住可根据需要选用。

虽然为配合业主的工期和成本控制，结构采用了较常见的混凝土框架-核心筒系统。但针对本项目的需要，采用了一些结构措施。为配合建筑的形态，在建筑曲线形的立面，采用了斜柱方案；同时为了提高建筑平面的效率，在塔楼的中低区使用了劲性柱以减小柱截面。

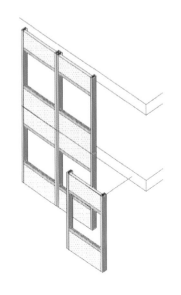

Curtain
幕墙

Fire Prevention for Glazed Wall:
1. Hollow or doubling hollow low-E glass
2. Horizontal material
3. Stone decoration strip
4. 1.5mm galvanised steel sheet
   (smoke prevention)
5. Fireproofing glue
6. >800mm fireproofing insulation
   in order to prevent flame rising
7. 3mm aluminium rear panel
8. Single pane glass
9. 100mm fireproofing/insulation cotton
10. 1.5mm galvanised steel sheet
11. Aluminium alloy louvres

玻璃幕墙防火措施:
1. 中空或夹胶中空双银Low-E玻璃
2. 横料
3. 石材装饰条
4. 1.5毫米镀锌钢板（防烟措施）
5. 防火胶
6. >800毫米防火隔离阻止火焰蹿升
7. 3毫米铝背板
8. 单片玻璃
9. 100厚防火/保温棉
10. 1.5毫米镀锌钢板（涂防火漆）
11. 铝合金百叶

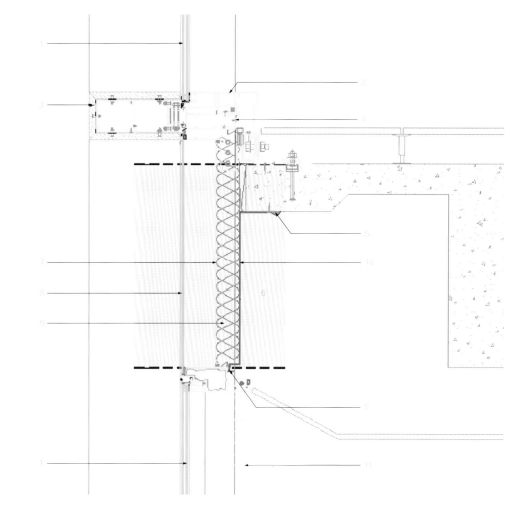

Glazed Wall's Energy-efficient Strategy:
1. Hollow or doubling hollow low-E glass
2. Appropriately increasing
   the air thickness to 12mm
3. Sectional materials with insulation materials
   inside and outside (Nylon 66)

玻璃幕墙节能措施：
1. 中空或夹胶中空双银Low-E玻璃
2. 适当增加玻璃间空气的厚度至12毫米
3. 型材的室内外间设置隔热材料（尼龙66）

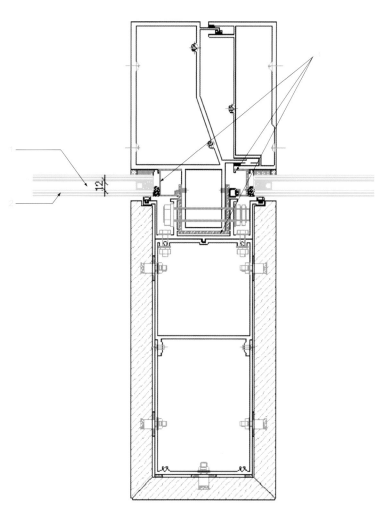

7. Lobby
8. Level 1 lift lobby
7. 大厅
8. 一楼升降机大堂

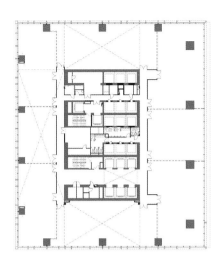
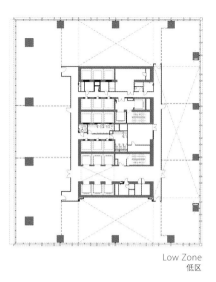

Low Zone
低区

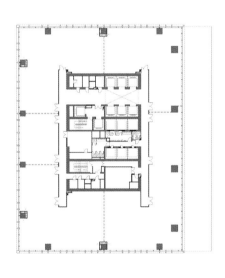
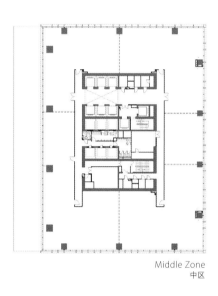

Middle Zone
中区

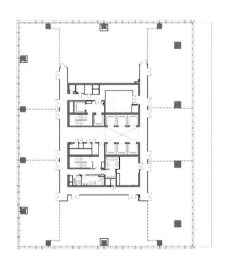
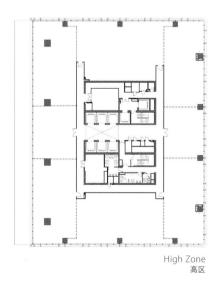

High Zone
高区

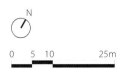

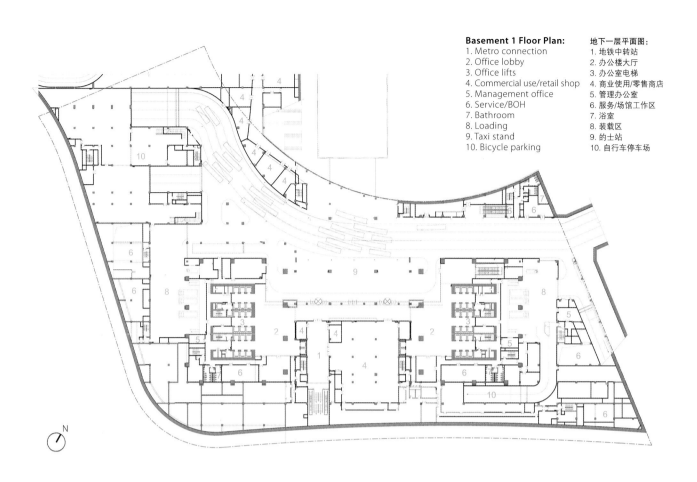

Basement 1 Floor Plan: 地下一层平面图：
1. Metro connection — 1. 地铁中转站
2. Office lobby — 2. 办公楼大厅
3. Office lifts — 3. 办公室电梯
4. Commercial use/retail shop — 4. 商业使用/零售商店
5. Management office — 5. 管理办公室
6. Service/BOH — 6. 服务/场馆工作区
7. Bathroom — 7. 浴室
8. Loading — 8. 装载区
9. Taxi stand — 9. 的士站
10. Bicycle parking — 10. 自行车停车场

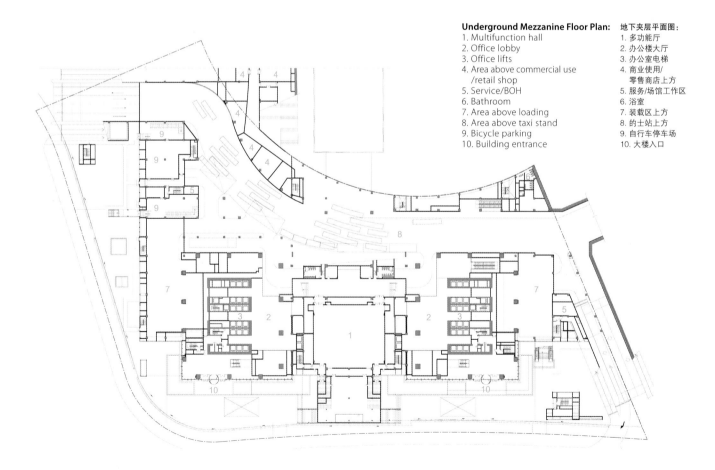

Underground Mezzanine Floor Plan: 地下夹层平面图：
1. Multifunction hall — 1. 多功能厅
2. Office lobby — 2. 办公楼大厅
3. Office lifts — 3. 办公室电梯
4. Area above commercial use/retail shop — 4. 商业使用/零售商店上方
5. Service/BOH — 5. 服务/场馆工作区
6. Bathroom — 6. 浴室
7. Area above loading — 7. 装载区上方
8. Area above taxi stand — 8. 的士站上方
9. Bicycle parking — 9. 自行车停车场
10. Building entrance — 10. 大楼入口

Lever 1 (Ground Floor) Floor Plan:　一层平面图：
1. Multifunction hall　　　1. 多功能厅
2. Office lobby　　　　　　2. 办公楼大厅
3. Office lifts　　　　　　3. 办公室电梯
4. Commercial use/retail shop　4. 商业使用/零售商店
5. VIP room　　　　　　　　5. VIP房间
6. Service/changing room　6. 服务间/更衣室
7. Bathroom　　　　　　　　7. 浴室
8. Building entrance　　　8. 大楼入口

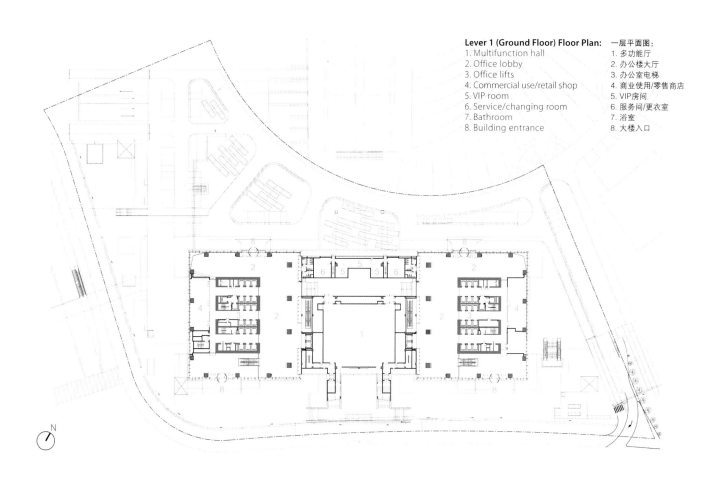

Level 2 Floor Plan:　二层平面图：
1. Skylight/roof　　　　　1. 天窗/屋顶
2. Office lobby　　　　　　2. 办公楼大厅
3. Commercial use /retail shop　3. 商业使用/零售商店
4. Office lifts　　　　　　4. 办公室电梯
5. Bathroom　　　　　　　　5. 浴室

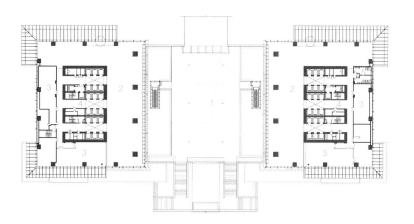

INDEX
索　引

012 / 022	马里奥·库西尼拉建筑事务所 **Mario Cucinella Architects** www.mcarchitects.it	134	王维仁建筑设计研究室 **Wang Weijen Architecture** wwjarchitecture.com
034	综合设计事务所 **Integrated Design Associates Ltd.** www.ida-hk.com	146	比尔·邓斯特 **Bill Dunster** www.zedfactory.com
044 / 084	苏州设计研究院股份有限公司 **Suzhou Institute of Architectural Design Co., Ltd.** www.siad-c.com	156	南京大学建筑与城规学院集筑建筑工作室 **IA studio, Institute of Architecture and Urban Planning of Nanjing University** arch.nju.edu.cn/index.htm
054	罗昂建筑设计咨询有限公司 **logon** www.logon-architecture.com	166	RMJM 建筑事务所 **RMJM** www.rmjm.com
066	刘宇扬建筑事务所 **Atelier Liu Yuyang Architects** www.alya.cn	176	HASSELL 建筑事务所 **HASSELL** www.hassellstudio.com
076	帕金斯威尔建筑师事务所 **Perkins+Will** www.perkinswill.com	184	思邦建筑设计咨询有限公司 **SPARCH** www.Sparchasia.Com
102	维思平建筑设计事务所 **WSP** www.wsp.com.cn	194 / 214	斯蒂文·霍尔建筑事务所 **Steven Holl Architects** www.stevenholl.com
110	欧华尔顾问有限公司 **The Oval Partnership Limited** www.ovalpartnership.com	226	KPF 建筑师事务所 **Kohn Pedersen Fox Associates PC** www.kpf.com
120	直向建筑 + 中建国际 **Vector Architects + CCDI** www.vectorarchitects.com www.chinaconstruction.com	234	TFP 建筑设计事务所 **TFP Farrellsi** www.tfpfarrells.com
		244	Arquitectonica 建筑事务所 **Arquitectonica** www.arquitectonica.com